RUNNER'S WORLD®

THE RUNNER'S BRAIN

ALSO BY DR. JEFF BROWN

The Winner's Brain (with Mark Fenske and Liz Neporent)

Chicken Soup for the Soul: Say Goodbye to Stress (with Liz Neporent)

Chicken Soup for the Soul: Think Positive for Great Health

RUNNER'S WORLD®

THE RUNNER'S BRAIN

HOW TO THINK SMARTER TO RUN BETTER

DR. JEFF BROWN
with LIZ NEPORENT

FOREWORD BY MEB KEFLEZIGHI

RODALE.

Rodale books may be purchased for business or promotional use or for special sales. For information, please write to:
Special Markets Department, Rodale Inc., 733 Third Avenue, New York, NY 10017.

Printed in the United States of America

Rodale Inc. makes every effort to use acid-free ♾, recycled paper ♲.

Book design by Amy C. King

Library of Congress Cataloging-in-Publication Data is on file with the publisher.

ISBN-13: 978-1-62336-347-5 paperback

Distributed to the trade by Macmillan

2 4 6 8 10 9 7 5 3 1 paperback

We inspire and enable people to improve their lives and the world around them.
rodalebooks.com

No one I know captured the joy, self-confidence,
and challenge of running more than my happy and clever son, Grant,
who is now running on streets of gold.

Grant Fieldon Brown

August 31, 2006–July 23, 2013

www.RunningonStreetsofGold.com

CONTENTS

PART 4: CHALLENGES

PART 5: RESOURCES FOR RUNNERS

FOREWORD

As a city, Boston is unmistakably accustomed to rallying for major sporting events. Whether it's a comeback victory in the World Series, an epic battle for the Stanley Cup, or yet another Super Bowl run, Boston gets it done. But I doubt that any sporting event in Boston's history—perhaps in the world's history—has been more anticipated than the 118th running of the Boston Marathon in 2014. It was that marathon that personified the words *Boston Strong* and defined our resilience and tenacity as a unified country of runners committed to a goal much bigger than ourselves. We chose to face fear, not run away from it. We chose to let our collective spirit be clearly seen by the world. We wanted to define who we are, what we do, and why we do it—bombs or no bombs—on our own terms. Even though I was blessed to cross the finish line first in 2014, in my heart, I believe we all won the marathon together.

The 2013 Boston bombing was something that touched all of us in the world of running. Whether you were a champion marathoner, a rising track star in high school or college, a parent who runs to set a healthy example for your kids, or a seasoned runner who just never kicked the decades-old habit of wearing down the soles on your shoes, all runners were affected by the bombing. And for those families, runners, spectators, and volunteers who have been affected permanently by the bombing, I continue to keep you in my prayers.

As human beings, we have thoughts, emotions, coping strategies, faith, and the ability to deal with obstacles. God gave each of us a brain that we sometimes forget to tap into when we run. We get distracted by the latest technology, our favorite brands, the hottest colors, the coolest clothes, and forget about our brains. As Dr. Jeff Brown says, "You already own the most high-tech piece of gear you'll ever need, and it's already paid for, too. It's your brain." He's right. No machine will function as fast and accomplish so much at once as your brain can.

One interesting and funny statistic to think about that Dr. Brown shares is that a Boston Marathon field featuring 30,000 runners brings approximately 90,000 pounds of brains out onto the course. Quick math translates that to 45 tons of grey matter, and that's something we can't ignore. We've focused on diet, stretching, breathing, long runs, short runs, fun runs, arch supports, muscle repair, hydration, blisters, and bleeding nipples. It's now time to focus on our brains and keep them permanently in the spotlight.

During Dr. Brown's 15 years of being the official lead psychologist for the Boston Marathon medical team, he's seen most everything mental that a runner can experience—and then some. He understands the unique qualities, special circumstances, deep motivations, and the confident joy of accomplishment through hard work. When I crossed the finish line in 2014, Dr. Brown was right there. My wife, Yordanos, grabbed him in a bear hug as she started jumping up and down, shouting "Praise the Lord!" Caught up in the moment, he jumped and shouted with her.

Dr. Brown's knowledge and experience with and about runners and their brains is vital for each of us as we, ourselves, seek to better understand how to run smarter and think like the healthy, knowledgeable runners we need to be. Just think what each of us might be able to accomplish if we committed to training and caring for our brains just like we do all of our other running goals. *The Runner's Brain* is full of solid science, proven mental strategies, inspiring and funny stories, tips, and thoughts from some of the greatest brains in running and is written in a down-to-earth, practical, and approachable style. Any runner looking to maximize his or her performance and get the most out of every run can benefit from Dr. Brown's wisdom.

Meb Keflezighi
June, 2015
San Diego

PART 1

RUNNING AND YOUR BRAIN

AS A RUNNER, your biggest asset (or sometimes your greatest enemy) is your brain. You must put in the miles and the time to run your best, of course. But I believe physical conditioning alone isn't enough to put wings on your feet. What you think and feel on and off the road has a huge influence over how you perform once you lace up.

Being the head psychologist for the Boston Marathon for the past 14 years and having worked with hundreds if not thousands of runners, I can tell you that understanding the mechanisms behind your thoughts and emotions can help you get to the top of your game. You meticulously plan your speed days and hill work; the mental aspects of your program should get the same priority and attention.

In this section, I'm going to dive into the theory of why running is good for the brain and why the brain is good for running. I'm going to explain how the neural wires and circuitry that run through your gray matter can enhance your self-image as a runner, and how you can fine-tune those connections.

I'll also take you behind the scenes at the 2013 Boston Marathon to give you my firsthand account of what happened when the bombs went off. That the attack on Boston affected every single American goes without saying. But I think for those of us in the running community the feelings run deep, even those who will never set foot on the Boston course. By virtue of being a runner, you have a personal connection to this national tragedy that informs your thinking and influences your emotions. So let's start off with that.

CHAPTER 1

The Medical Tent

Lessons from the Boston Marathon bombing

SO A PSYCHOLOGIST walks into a medical tent and the bartender says, "What's a psychologist doing in a marathon medical tent?"

Have you ever heard this pitiful joke? I have. Dozens of times. Except that most of the time people are asking it as a serious question. As in: "Seriously, how did a psychologist wind up in the medical tent of the Boston Marathon?" Actually it's a funny story.

I came to Boston in the late 1990s to do a postdoctoral fellowship in child and adolescent psychology at Harvard Medical School, specifically McLean Hospital. My interest in sport psychology had traveled with me from Louisiana where I did some consulting for a minor league baseball team and a local college athletic department. There I developed a passion for helping athletes enhance their performances by enhancing their brains' capabilities. Athletes are quick learners once they put their minds to it—no pun intended.

I specialize in a solidly research-based area of psychology known as cognitive behavioral therapy, or CBT. This is a type of psychological approach where you focus on negative thinking patterns and challenge an individual to recalibrate and restructure the brain with more productive thoughts. In turn, performance can be affected in a positive direction. Research reveals that CBT makes permanent alterations to the brain. So it's the obvious treatment of choice for training the brain and changing behavior. It helps focus on the accuracy and content of your thinking, thus rewiring your brain in a positive way. CBT can offer incredibly powerful tools for athletes in general and, as I have discovered over the years, runners specifically.

I hadn't been in Boston long before I met my colleague Dr. Arthur Siegel, who is the director of internal medicine at McLean Hospital and, by chance, a runner and a long-standing member of the marathon medical team. One of Siegel's clinical research interests is hyponatremia, a condition a runner acquires when his sodium level becomes too low, usually from drinking a vast amount of water that doesn't clear his system fast enough. Many of the first signs and symptoms of hyponatremia are psychological in nature.

In chatting about this, Siegel kept saying how a psychologist could be useful on the marathon medical team to help spot hyponatremic runners for treatment by picking up on the psychological symptoms. Those conversations ultimately spun themselves into an invitation to become a member of the marathon's official medical team.

At the time, this was a unique position—and in Siegel's words, I was like a canary in a coal mine because of my ability to quickly detect the psychological stress of a runner that often accompanies hyponatremia the way a canary can quickly detect dangerous gases in very small amounts.

As an aside, I would like to point out how this demonstrates just how thoughtful the medical team leaders and clinicians are about the care provided to runners on Marathon Monday. The medical tent doesn't happen just overnight; the thousands of volunteers who help out don't just show up by instinct. Runners who find their way to the medical tent meet a response that has been planned, reviewed, and fine-tuned each and every year. It's a privilege to work with this team of professionals.

Part science, part gut instinct, the canary role quickly grew into something more. I discovered that just like a coal mine, a runner's brain, too, has multiple caverns, weight-bearing beams, fault zones, and upcast shafts that contribute to its structural function and integrity. Only rather than a coal mine, I found the runner's brain a gold mine, filled with extraordinary abilities, an amazing capacity to learn, and a vast hunger for knowledge.

The medical tent at Boston, and at any other big race, is in place to treat runners with more serious conditions that need immediate medical support. Those who want bags of ice, gauze for their bleeding nipples, or a needle to drain a patch of blisters usually don't require the services of the tent; it's more appropriate for runners who have collapsed at the finish line or were pulled off the course for abnormal core temperature, severe cramping, delirium, mental confusion, cardiac concerns, or any of a host of other critical medical events. Some of these problems crop up during the race, but sometimes they are a result of a condition that existed prior to the race that may have been exacerbated by any number of factors.

Initially my job was to recognize the psychological signs of hyponatremia more quickly than lab work can confirm it. I was to alert the physicians on the team to the potential problem so the runner could be treated appropriately. I was then to continue to help the runner deal with the experience from a psychological and emotional standpoint until the appropriate treatment started working.

While I was on the lookout for hyponatremic runners that first year, it became clear to me that while my diagnostic skills were useful, my CBT training could be even more so. Runners who were experiencing pain of various kinds, dealing with anxiety or anger related to their performance, or feeling depressed and disappointed could benefit from altering their negative thinking. The way I saw it, runners were fanatical at avoiding the physical injuries that came from pounding the pavement but rarely planned for mental maladies. I've always believed runners should be treated for psychological woes in the same way they are treated for a pulled muscle or an achy knee.

By the time 2013 rolled around, I had been the psychologist for the Boston Marathon for more than a decade. I thought I'd seen it all—or at least, close to it. That's why the book you hold in your hands came to be. I wanted to help runners understand, use, and strengthen their brains. I had treated every psychological symptom you could possibly imagine that is associated with running: runners who had forgotten to take their psych meds only to experience a resurgence of psychiatric symptoms in the middle of a race; the psychological effects of hyperthermia, hypothermia, and low blood sugar; runners who had collapsed due to event-related anxiety; competitors confused and worried about losing their composure during a panic attack; those who needed an IV being petrified of needles; individuals who had succumbed to a quick-onset depression because of a poor performance; and so much more.

I thought I was prepared for any psychological or emotional problem the marathon could throw a runner's way. Then, the bombings.

WHAT HAPPENED THAT DAY

Medical Tent A is an expansive, rectangular tent situated across the entire four-lane street in front of the Boston Public Library in Copley Square. It sits about 100 yards from the finish line on Boylston Street, one of the prettiest and most populated streets in downtown Boston. Across the way sits Tent B, a mirror image of Tent A.

In 2012—just as it had been for every year before—Tent A was crowded, with nearly 200 shiny, Mylar-covered cots waiting to receive runners needing attention. Ropes were stretched taut along the length of the tent for hanging IVs, and a blood lab was centrally located for quick turnaround. The tent was a busy, buzzing hive of doctors, nurses, podiatrists, physical therapists, cardiac specialists, a family resource team, translators, scribes for medical charts, and communication experts—all at the ready to receive runners in distress as they entered for medical treatment.

As I recall so vividly from that afternoon, I was working with a runner in pain on some dissociative strategies related to severe cramping in his legs. My face was probably about 2 feet from his face when I heard the first bomb go off.

I knew right away that it wasn't a sound we usually hear in New England. We have cannons. We have fireworks. Occasionally you see a news story about a manhole cover blowing. This was a completely different sound because it was followed by a kind of a hollow echo—and because I was hearing it on Boylston Street.

I made eye contact with a physician friend of mine across the aisle, who was working with a different runner. We immediately understood that something was wrong but tried to appear as unruffled as we could, continuing what we were doing because that's what you have to do when you have a runner right in front of you who needs help.

A second blast happened just seconds after the first. This was the second bomb. Another one of my colleagues approached me and told me calmly—without yelling or showing too much emotion—that there had been a bomb at the finish line. At that point the physicians were asked to go to the scene and assist Boston's emergency medical team as first responders.

I headed out of the tent to distribute protective gloves because I noticed most people had run out without remembering to grab gloves, which are so important for sanitary and safety reasons. I wanted to play my part by making sure everyone had what they needed. But rather quickly, the injured started gathering back in the medical tent, so that's where I went, too.

I stayed in my role as the race psychologist by trying to divert the less injured away from the more critically injured, hoping to avoid the even greater potential trauma of their witnessing victims who were obviously in deep medical trouble.

A lot of the runners who came into the tent had been separated from their families and were worried about their loved ones. I recall speaking with one family whose four members had been injured—one fairly significantly—but they were all able to talk. At one point someone else was brought in on a gurney in really bad shape, and naturally they started to turn and look. I told them to keep their eyes on my face, and they did—they turned back around and looked at me. They were already going through so much. I tried to reduce the level of what they had to deal with.

Many people in the tent, both the runners we were treating and the volunteers, were terrified and trying to make sense of what had happened. They wanted explanations. I tried to reassure those who I interacted with that they were now safe; I kept telling them that what happened should not have happened, and that I was very sorry they had to go through something like this.

In a mass casualty event like what happened that day, you realize you can only do so much. But as a medical team, we were able to do so much more than any one individual. We saw a wide range of injuries come into the tent, from shrapnel wounds to people who had lost limbs. It's a lot for anyone to process. My goal was to try to help people I worked with handle everything the best they could. A lot of them were asking why: Why did this happen? Who did this? These are normal questions that come up, and in real time we didn't have any answers that made sense. Whatever the answer was, though, it would be malicious and deliberate.

Immediately after the bombings my job was to reassure every runner I spoke with that no, this is not normal and, knowing that a bombing and related injuries aren't normal, it's okay to wonder about the how and why of it all. It's what everyone is asking themselves and each other. It's perfectly appropriate to ask those questions.

That's really all I want to say about the 2013 Boston Marathon bombing. I don't wish to recount the blood and gore of that day. That would be disrespectful to the runners, spectators, and everyone working the medical tent. I don't believe the bombers were targeting runners or their fans. They were targeting a symbol of the American way of life, not a specific person or runner. That's why they didn't just hit the grocery store down on the corner in the neighborhood: It would never make worldwide news. But the World Trade Center, the Pentagon, the Boston Marathon—these are all symbols of freedom, America, international camaraderie, strength. It's better to concentrate on the amazing response of the volunteers who literally saved life and limb for so many that day.

HEALING HEARTBREAK HILL

The reason I wanted to talk about this experience at all is because I believe it is a shared experience for all Americans and people all over the world, but especially the running community. For anyone who loves the sport of running—whether they will ever run Boston or not—this was a deeply personal event. It was painful for those of us who enjoy the simple freedom of putting one foot in front of the other to see others who do the same put in harm's way.

Boston Marathon organizers acknowledged this emotional dilemma. That's why the following year our leadership team upped the marathon's usual psychological support staff from 2 to 60. In 2014 we had mental health professionals in 22 of the 26 medical tents along the route, in the chutes at the finish line, and in both of the race's main medical tents. We also had a private area at the runner's expo in case anyone wanted to come in and talk about anything from race strategy to how they were processing the previous year's events.

I am proud of the running community's incredible ability to heal and find the good and support resilience in each other, and for the overwhelming support they continue to give to the injured runners and their families and the community at large. It demonstrates the spirit of runners, and their kindheartedness. Though it may change how some of us prepare psychologically for an upcoming event, we don't have to be stuck in the trauma and shock forever.

As for me personally, I think with the passing of time and putting everything I witnessed into the perspective I could, this experience has helped me become a better psychologist to runners and our medical team. I think many of them, especially those who run in Boston and in other big marathons, find it helpful when I can tell them I was there, that I saw what happened. It seems to set a boundary for them: If the person talking to them has dealt with it, then they can deal with it. People feel they can tell me exactly what's on their minds because in some unfortunate way, having been there lends me some credibility. It allows me to build a quicker, stronger alliance with runners.

CULTIVATING A "RUNNER'S BRAIN" MENTALITY

With the bombings, we as runners were asked to deal with a long road beyond the 26.2 miles we signed up for. Even if you're just a jogger who has never competed or a casual racer who likes to run in local charity 5Ks, those bombings in Boston touched you and made you think about safety. That's what terrorism is about; the aim of such malice is to make you wonder about your safety and to diminish something that's strong and powerful. It doesn't matter if you ran Boston in 2013 or not, or if you ever run any other race. What's important is that we don't let this experience stop us from doing anything as important as a daily run. So when you lace up, when you're out pounding the pavement in the rain, when you're out there in the early morning ahead of the school buses, those are times you are preserving what is important to you and to our country.

Runners have demonstrated clearly that the bombing hasn't stopped them. Fortitude is just one amazing trait runners possess. You can see somebody coming into the medical tent in a wheelchair, screaming because their legs are cramped up. After an hour they look completely different; they get up and walk out on their own, smiling and thanking the team that took care of them. For me, the bombings created a bigger version of that response. None of us can walk away from it in an hour, but I think we can, over time, come to terms with the experience.

Beyond the obvious lessons in rising to the occasion and showing resilience, what does the Boston experience have to do with having what I call a runner's brain? Everything, I think.

For runners, this marathon brought the human brain into the spotlight, right up there with the latest running shoes and high-tech gear. We are always talking about how to improve the physical aspects of running, but I think it's time that we start giving the brain its due. For a runner, the power of the brain is immense. It has the ability to impact your progress just as much as your quads and hamstrings. It guides your every step during a run; it monitors the good, the bad, and the pain; it tells you when something is off and problem-solves to correct it on the spot if it can. The capacity of the brain is limitless. It never stops, even when you feel like you have to.

Now, if it's okay with you, I would like to move past this experience and talk about how to use your brain for the good of your running program. As I always say, if you're going to be a runner, you've got to have a runner's brain.

CHAPTER 2

Five Ways Running Jogs Your Brain

What running can do for your brain

YOU ALREADY KNOW that running does a body good. Maybe you also know that running does some pretty terrific things for the brain, too. So terrific, in fact, that the benefits it bestows upon your mental matter make any physical improvements seem almost incidental.

Almost every single dimension of cognition can be enhanced by going for a run. Some benefits are instantaneous; you begin to experience them from the moment you break a sweat. Some help sharpen your mental skills temporarily, but are still very useful brain "superpowers." Others are the cumulative effect of a lifetime of running, the result of the brain gradually reshaping itself over the course of many runs.

In case you didn't know just how good your running habit is for psychological health, I thought I would start out by clueing you in. Because running does so many amazing things for the brain—too many to catalog, really—let's focus on three of the most spectacular, well-researched effects.

IT SHARPENS YOUR MEMORY

Running is especially good because it quite literally jogs your memory. One study,[1] carried out by scientists at the University of British Columbia, focused on a few dozen women in their seventies and eighties who tested positive for mild cognitive impairment, a condition that leaves a person's memory foggier than you would expect even for someone of advanced age. The scientists asked some of the women to jog or briskly walk on a treadmill several days a week, another group to lift weights, and a third group to do some light stretching and toning exercises. After six months, the scientists gave all the women a memory test and compared the results to tests the women took at the start of the experiment.

Women who stretched and toned lost some cognitive ground, with the tests revealing that their memories faded. Both the joggers and the weight lifters improved spatial memory, which stores information about places and where things are found. However, only the treadmillers made gains in the type of memory that helps you recall words and other aspects of language.

It's important here to underscore the significant take-home message from these findings: The joggers' and walkers' memories didn't just hold steady—their memories actually *improved*. If memory had simply resisted deterioration, that would have been a nice enough outcome. But gaining memory in your golden years? Well now, that's a gift.

This is only one example from a growing body of evidence showing how all kinds of cardiovascular exercise (much of it tested on runners and walkers) bulks up memory. It seems to do this by triggering growth of a protein known as BDNF, or brain-derived neurotrophic factor, which augments the health of existing neurons and encourages the creation of new ones, especially in various areas of the brain associated with memory creation. Put another way, running acts as a sort of Miracle-Gro for your memory. With each footfall, you not only preserve the memory that's already there, you may potentially expand and enhance it, too.

IT AGE-PROOFS YOUR BRAIN

From the moment you are born until the moment you die, your brain is constantly reshaping itself. All your thoughts, behaviors, emotions, and experiences are like an ever-growing crowd of sculptors that form the building materials of your gray matter into its amazing, distinctive form.

Running is a special kind of sculptor. Rather than leaving its fingerprints all over the brain, it leaves footprints. By increasing bloodflow to the brain, it encourages the biochemical tools of "neuroplasticity" to enrich the production of new connections between neurons and strengthens connections within the neurons themselves. At the same time, it bathes newborn neurons in BDNF so their functional capacity flourishes.

Proof of this phenomenon comes from, among other sources, work done at the Salk Institute for Biological Studies in San Diego.[2] The research team there found that physical exercise of any type can help generate new brain cells, even in the aging brain. Their work shows that in as little as three hours a week, sustained cardiovascular activity such as running halts and even reverses brain shrinkage, especially in the areas responsible for memory and higher cognition.

In other research, brain imaging tests have demonstrated fortified neural pathways and boosted cognitive flexibility even after just one jogging session. All this evidence suggests running is one of the best ways, if not *the* best way, to keep the brain young even as you age.

IT MAKES YOU HAPPY—AND SOMETIMES EVEN EUPHORIC

Of course, plenty of couch potatoes are happy as clams. But as a runner you actively set yourself up for contentment and peace of mind. Here's how.

Each run is an aspirational event. You set a goal that once met gives you a sense of accomplishment and pride. If you happen to lose weight and tone up in the process, well, that can make you feel pretty good about yourself, too. And as I'll explain in the chapters to come, running stimulates the mind and helps grease the skids of creative thinking so you are more engaged with the world around you. We're also going to discuss goal setting because I consider it one of the key Runner's Brain strategies.

Exercise promotes specific neuroplastic changes that boost mood and self-confidence, reduce anxiety, and promote relaxation. For starters, it amplifies the production of endorphins, the "get happy" chemicals that have been shown to reduce the perception of pain and lift the spirits. For another, it lowers the levels of the stress-associated hormone cortisol. At the same time, both exercise and positive emotions strengthen the immune system, so you're left feeling sick and miserable less often.

Scores of studies have tested the hypothesis that any type of aerobic activity can help melt away worry and put you in a good mood. One team of investigators had a third of 154 clinically depressed subjects take antidepressants, another third take part in a cardio exercise program, and the final third do both.[3] At the 4-month mark, more than 60 percent of the subjects in all groups were no longer considered clinically depressed. A follow-up study done 6 months after the original study ended found that the effects of exercise lasted far longer than the effects of the medication.

I'm not suggesting you trade in antidepressants prescribed by a doctor for running shoes. But I am suggesting that if you've been feeling down in the dumps, a few laps around the block certainly can't hurt and will very likely help. I find it ironic so many people skip their cardio workouts when they are depressed; it is probably the number one thing that can help lift their spirits, because it rebalances their brain chemicals. Of course, depression affects motivation and energy levels, and that's the rub for not seeing the green light. And if you are taking antidepressants, it's worth a discussion with your doctor about how regular exercise might promote a better response to them. Please ask her.

By the way, that runner's high you hear so much about is a special case of happiness. I've heard some runners say the experience makes their brains go all sparkly as this amazing sense of oneness between mind and body swells within them. Though the feelings are fleeting compared to the long-term anti-depressive effects I mentioned previously, they can have real, long-lasting effects on motivation. Even if you never experience a runner's high—and many lifelong runners do not—some level of those feelings usually find a way to shine through.

Scientists believe that, runner's high or no, the act of running stimulates the brain's reward centers including those located in the areas of the striatum and nucleus accumbens. When these centers catch the buzz from running once or twice, they may make you crave a run like some people crave chocolate or cigarettes. You might even say that when this happens, running can become slightly addictive.

We know that chemicals called endogenous opioids are released during intense physical activity. These chemicals, which are the body's naturally occurring versions of drugs like opium and morphine, bind to your brain receptors associated with pain management, pleasure, and relaxation. We also know that the brain releases its own form of endogenous cannabinoids, similar to the cannabinoid substances found in marijuana. In mouse studies, stimulation of the cannabinoid receptor sites on the brain has been associated with increased wheel-running. (Though it goes without saying that smoking cannabis in general is not the way to jump-start a running career.)

Although we know runner's high is possible, we don't know exactly why. There is no clear-cut theory why some people experience it and other feelings of happiness from a workout. Anecdotally, it seems newly minted runners are more primed to have these euphoric sensations than seasoned runners, so it might be neurobiology's way of telling you to push past the initial pain of shin splints and sore muscles and stick with it. Or maybe it's self-preservation, your brain's way of ensuring you become a lifelong runner. Whatever the reason, it's a great feeling for those lucky enough to have the occasional experience. In Chapter 7, I'll go into more detail about the phenomenon of the runner's high, how you might chase that state of bliss while running, and why it's okay if you never catch it.

WRAP-UP

Besides these three well-understood effects—sharpens your memory, age-proofs your brain, and makes you happy—running is linked to a host of other brain enhancements that are less well understood. I have no doubt the proof will continue to pour in about the brain benefits of running as research continues. Not that you need to justify your dedication to the leggy arts—clearly you're already a fan. But it's nice to be reminded of how much you will be rewarded for your efforts.

CHAPTER 3

I Think Therefore I Am a Runner

How the brain forms your identity as a runner

IMAGINE YOU ARE AT a noisy cocktail party. The packed room is so boisterous you can barely have a conversation with the other guests crowded around the crab dip. Suddenly, someone across the room mentions your name. The noise is snuffed out like a candle, and the crowd seems to part. As if by magic, you are able to filter out all distractions and train your ears on the source of the name-check.

Why does your brain do this? If someone is speaking about you, you probably want to pay attention. They might be complimenting you, talking about something you need to know, or attacking your reputation. You don't know unless you listen in.

But how does your brain do this? Anytime your brain shoves irrelevant data aside to focus attention on significant information (like your name) it owes it to a neural region called the reticular activating system.

RAS for short, this loose network of neurons and neural fibers originates at the brain stem, in the very back of your brain, and winds its way through the rest of your gray matter to help manage sleep, breathing, and heart rate. But its real superpower is sifting through incoming information to decide which bits get attention and which bits are ignored. This is really good news for new runners, seasoned runners, or anyone in between.

Think of your RAS as your brain's speed dial, programmed with all the important phone numbers. If it hears any of those numbers dialed—such as your name in the example of the cocktail party, or some novel feedback like the tickle of a spider crawling up your leg—it shuffles that information between your conscious and subconscious to ensure your brain is listening in on the call.

If it weren't for your brain's RAS circuitry, your consciousness would be deluged by an incoming sensory overload and you would have trouble sorting immediate considerations. By amplifying pertinent factors within your environment to the top of your attention, the RAS helps you prioritize and direct your concentration to where it's needed.

Say, for example, you are searching through your computer for one file in particular. On some level you sense everything around you: the hum of the air conditioner, your coworkers chatting, the feel of your fingertips on the keyboard, the dozens of file names that come up on your search. It's the job of the RAS to filter out all that background noise to allow you to scan for the relevant words without distraction. If the file you're looking for seems to jump right out at you, that's the RAS successfully doing its job.

WHAT THE RAS HAS TO DO WITH RUNNING

Perhaps this has happened to you: You're standing in line at a grocery store in your running tights and sneakers. You strike up an idle chat with the person in line behind you about the price of broccoli or what have you, and after some time, the person points to your outfit and asks if you're a runner.

"Well, you know, I run but I'm not really a runner," you might sputter.

Or you might say: "I run but I'm really, really slow."

Or: "I've done a couple races but I wouldn't consider myself a runner."

Francie Larrieu Smith told me that she sometimes doesn't know how to respond when someone introduces her as a runner—even though she's a five-time track and field Olympian and a successful coach who considers herself a lifelong runner.

Before you build the confidence to respond with a "heck yeah I'm a runner," your RAS needs to believe. And it needs to send a message to the rest of your brain that it, too, should believe. Your job is to throw the RAS as much and as many different types of information you can about your being a runner. This is how you can strengthen your identity as a runner.

It works like this: The RAS influences so many different parts of who you are. One more important piece the RAS influences heavily is cognitive function, which in turn directly influences your belief system. One of its most essential jobs is to spot, select, and retain any information that supports your view of the world and of yourself. Think about everything you've experienced in your life related to running: the first pair of running shoes you bought, the first time you did speed work, the first time you ran a race, the first time you ran 10 miles.

All of these "firsts" were probably pretty engaging as they were happening. You likely felt a high level of enthusiasm because these occurrences were fresh and interesting. But after a while those flutters of excitement became less frequent. Now these aspects of running are woven into the fabric of your identity as you buy shoes routinely, run track workouts twice a week, enter races all the time, and do a weekend 10-miler every month.

The building of these routines and rituals happen by instinct because you have formed a belief system. Negative thoughts undermine the belief system, while positive ones reinforce it. So if you are able to strengthen your beliefs by filling up the RAS with constructive, confidence-building experiences and information, you start to feel and act the part of a runner.

The greater the number of reinforcing episodes, the more powerful these beliefs become. And by the way, the RAS can only hold so much information, so that's why you want to deliberately send "I'm a runner!" messages to it over a lengthy period of time.

Now, if you have doubts or feel you don't deserve to call yourself a runner, you'll look for all sorts of evidence to support that idea: You're too slow; your body isn't the right shape; you don't run enough. Your RAS can be tempted to accept all of these thoughts as irrefutable proof you're not worthy to call yourself a runner. Often runners tell me they believe all the negatives and setbacks because they haven't achieved much with their running yet. Usually it works the opposite way: You believe first, then you achieve. And remember, the RAS can only hold so much information, so choose that information wisely.

You need to send very specific cues of your goals and identity-to-be to your conscious mind to solidify your confidence. The RAS will pass these notes on to your subconscious, and a cycle of conscious to subconscious and back again will fortify the information you want to preserve. So when you believe you are a runner, you become one. This will remain true despite evidence to the contrary that can sometimes happen even to the best of us.

Say, for example, you've been injured for a couple of weeks and haven't been able to run. You may start to feel discouraged and possibly consider giving up on a running career. But if your brain's RAS is bursting with affirmative thoughts, the right environmental cues, and secure behaviors, your identity will be much less likely to waiver. Rather than giving up, your RAS will help you make it through tough times and pick right back up as soon as your body is ready.

I had a talk recently with Amby Burfoot, a Boston Marathon winner and editor-at-large for *Runner's World,* that perfectly summed up why you shouldn't let setbacks dictate how you feel about yourself as a runner. As he points out, when he was competing he didn't win every race and he didn't set a world record every time out.

"I had more good days than bad days," he says. "The trick is to keep the good days in the front of your mind and remember that your achievements aren't a fluke." According to Amby, your best days represent your potential, and that's often reason enough to keep running. I think this is wise advice.

So when you're asked the question, do you hesitate to admit you're a runner? If so, your RAS needs a pep talk. Let's flood it with all the right thoughts and fill it up with proof that you've earned the right to that identity. But remember, even a hard-working editor like the RAS can only process so much before hitting overload. That's why you must be selective and intentional about the neural text messages you send it. Satisfied, successful runners of all levels deliberately fill their RAS with good vibrations. Let's go through some exercises that nudge your RAS in the direction you want it to go.

Runner's Identity Strategy #1: **Stuff Your RAS Full of Knowledge**

You've heard the expression knowledge is power? In the case of the RAS, knowledge is belief. You want to learn everything you can about running and being a runner. So here's my plug for buying that subscription to *Runner's World,* and any other running-related reading material you can get your hands on.

After reading them, lay them where you can see them, or cut out special articles or pictures and post them at work. Comb through this book, pour through the magazines, and surf the running sites on the Internet. Consider starting your own running blog, just for friends and family to read—they are some of your biggest fans and will reinforce that you are a runner when they post comments or compliment your writing. Study up on everything from technique to race-day strategies. Don't just read the how-to's either; you can glean quite a lot from biographies. It's one of the best ways to learn from others, often in their own words, exactly how they achieved their goals—literally step by step. (For a taste of the lessons you can get from biographies, turn to Chapter 20. Some of the greatest runners ever tell it like it is and offer their sage advice on the mental side of the sport.)

Runner's Identity Strategy #2: **Hang with Runners**

They're doing loops of the park every morning. They're attending clinics. They're on the message boards. You know where to go to find the people who can teach you the lingo and the tricks of the trade. Think of other runners, particularly seasoned runners, as the wise men and women of the tribe. Most runners I know are more than happy to share their war stories with you. By hearing what they have to say, and having the opportunity to ask questions of someone who has been down the road you are traveling, you can learn a trick or two and avoid a trap or two. If you're lucky, you will find the perfect running group and perhaps a mentor to take you under his or her wing.

I like marathon great Jeff Galloway's motivation for how he became a runner. It's the perfect illustration of runners influencing runners:

"I was a fat kid when I started running. My eighth grade requirement was that every boy go out for strenuous athletics after school.

"I had gotten to know some of the other lazy kids, and they said that winter cross-country was the one to do because you could lie to the coach and tell him you were going to run on the trails and you could go out to the edge of the woods and hide out. I did that for two days and then an older kid that I liked came up to me and said: 'Galloway, you're running with us today.'

"I guess I'd been busted. I ran with them and I was going to drop out at the woods but they were funny. And they told interesting stories and gossiped about the teachers as they ran. I stayed as long as I could but I couldn't run very far that first day. I got hooked during those 10 weeks on the social aspect of running and the way I felt empowered."

By the way, even if you yourself are one of the wise ones, you can still learn a lot from other runners. Not just the veterans, either: Newbies may not have much in-depth information to impart, but they bring a fresh perspective to the table.

Runner's Identity Strategy #3: **Program Your Self-Talk**

The RAS is always on the alert for relevant information. It will allow significant data from any of your senses—tactile, auditory, visual—to interrupt whatever you're doing if it's relevant to survival, safety, or your predetermined goals. That said, you can also purposefully manage sensory gates through choice and intention. Priming your brain with positive affirmations is a good way to stay motivated and feel successful. I'm going to explore the idea of self-talk in many of the upcoming chapters.

Note that your RAS can't tell the difference between a real event and an imagined one. It simply gives attention to any message you send it. When you imagine, for example, that you can run 5 miles with ease, this pretend practice may help improve your actual ability to run that distance. While visualizing an event is certainly no substitute for the real thing, it comes very close. As you will learn in upcoming chapters, it's a powerful pairing to combine your physical training and brain power to make you a better runner.

Runner's Identity Strategy #4: **Set Goals**

I hit the idea of goal setting hard in this book. Why? Because I believe that goal setting is probably the most valuable tool you have for achieving anything you set out to do as a runner, whether you want to compete in a local 5K or win an Olympic race. Even when you don't realize you're thinking about these goals, your brain knows that they're important and makes note of anything that might relate to them.

Story after story from some of the top runners in the world has convinced me that every single one of them had a laserlike focus on their goals, whether they consciously realized it or not.

Ultramarathoner Dean Karnazes is known as "the man who can run forever" because he can run for three days straight with little sleep and feel almost no fatigue. He told me he is constantly setting goals.

"Goals, to me, are short-term milestones and are an endless procession of ongoing 'baby steps' on the path to reaching a dream," he says, adding that he starts with a dream and works backward in aligning goals that will help get him there. Dean is right on track with how to set goals.

"My next dream is to embark upon a worldwide expedition to complete a marathon in every country of the globe in a 1-year period. There are 204 countries and I'm working with the State Department and the UN to get the necessary passports and permits to be able to do this. As you can imagine, the planning and logistics are every bit as complex and difficult as the running itself. Using goals to complete all the necessary tasks is an indispensable tool in the process of making this dream come true."

You don't have to be elite or even a natural runner for goal setting to help develop an upbeat RAS. I've seen average runners have extremely satisfying careers by keeping their goals front of mind. Goals help you build a positive sense of self-confidence by clearly defining success. You cannot be successful unless you recognize success when you see it.

Runner's Identity Strategy #5: **Dress the Part**

One way to feel more like a runner is to dress like one. Dressing the part to feel the part is a theory known as "enclothed cognition." Suffice it to say, for now, that by wearing a professional-looking fleece, sleek tights, and a brightly colored pair of kicks, you look fast—and this probably primes your brain to make you go faster. (I remember that when I was a kid, I could always run faster in new tennis shoes.) I go into this theory in depth in Chapter 9.

Runner's Identity Strategy #6: **Run!**

As obvious as it sounds: run, run some more, and then run some more. Run especially on the days when maybe you don't feel like it. Get into a routine, on a schedule, into the habit. The more you run, the more you will think like a runner. The more you think like a runner, the more you will believe you are a runner and the more you'll run. It makes perfect sense.

WRAP-UP

To become a runner, you have to learn how to think like a runner. You can teach your RAS to push irrelevant information aside and flood your brain with positively reinforcing messages. Consider starting with one runner's identity strategy, trying it out for a week or two and then adding on to that. You will boost your confidence, and I'm betting you will also get a bump up in performance as well.

PART 2

BRAIN STRATEGIES

THOUSANDS OF THOUGHTS pass through your mind every day. As a runner, presumably a percentage of those thoughts are related to running. Perhaps up until now your thoughts have been random and unplanned. I think that is the case with a lot of runners. Occasionally I will come across a runner who naturally thinks in a way that is most advantageous to performance, but most runners I have worked with can use a little help shaping their thought patterns to their advantage.

The first step to harnessing the power of thoughts, feelings, and emotions is to become more aware of them. In the first two chapters in this section, I'll enlighten you about just that. I'll also give you some guidance on how to nudge your thoughts in the right direction to make you more successful as a runner. After that, I'll explain why you probably don't have intentional control over higher states such as runner's high and flow, but I will show you how you can prime your brain so that you are more likely to experience them. Finally I'll tell you why we runners are so prone to magical thinking, and how you can use your superstitions to your advantage.

CHAPTER 4

Goal Setting

Infusing your runs with purpose

OLYMPIC MARATHONER AND expert running coach Jeff Galloway told me a story about qualifying for Boston the year after the bombing. He felt driven to meet the goal of qualifying in a way he hadn't felt since his competitive days, when he made the 1972 Olympic team. Aiming for Boston became a mission.

"During a clinic in my Philadelphia store I was asked about my feelings during the Boston bombing. It suddenly hit me that it was déjà vu: Munich. In 1972, a PLO group calling themselves Black September took Israeli Olympians, coaches, and staff hostage in the athlete's village and eventually killed them. We had several soul-searching meetings in Olympic Village about the terrorist incident there. The overwhelming meaning of this was the same as after Boston: 'They cannot take this away from us.'" (Four days after the massacre, American Frank Shorter won the Olympic Marathon. He knew that he was "exposed" on the course, but has said in interviews that once he started running he never thought about his safety because he knew that if he did, the terrorists would have won.)

So Galloway trained throughout the summer, got in better shape, then gave it a shot. He entered the Air Force marathon in Ohio and gave it a really good go. He came up 38 seconds shy of qualifying.

No one would have blamed him if he shrugged his shoulders and walked away. It was a valiant effort; he came up just short. Rather than feeling defeated, however, his failure helped him refocus.

"I started to look back on each race to see how they turned out, and most of them didn't turn out that well or what I thought would happen. That's when I started zeroing in on realistic projections," he recalls.

Galloway says that every single time out he made mistakes and learned from them. He recalibrated, kept his eye on the prize. He kept going. He adjusted his training. He rethought his workouts. When he tried his next qualifier a few months later, at the Space Coast Marathon in Florida, he made it—by 6 seconds.

"It was a tough day," he says. "It was a hot 70 degrees with a headwind, high humidity. Very tough at the end. I struggled during the last 3 miles and had to give it all I had during the last mile. It was the most satisfying running accomplishment in the last 20 years."

THE IMPORTANCE OF GOALS

In my office, I have a marathon bib with words written on it by Bill Rodgers: "Know why you are there." Just like Rodgers values goals, Galloway says he has realized the importance of having goals, even during periods he has run casually. He has imparted this ethic on the thousands of runners he has trained, and through the millions of books he has sold. Like me, he believes that goal setting is the foundation for every runner's successful mental strategy and runs.

So many runners I've talked to over the years fail to take stock of what they are truly capable of right now, or of what they might realistically be capable of in the future. They give absolutely zero thought to what they hope to accomplish through running. I think that's a colossal, unfortunate missed opportunity.

In my opinion, every runner should set goals. It can't be an accident that striving toward your objectives is one of the largest tasks assigned to the prefrontal lobe, the brain's seat of executive function and complex cognitive capabilities. Without goals there is no direction, no reasonable way to channel your energy. As the saying goes: "If you don't know where you're going, you won't find a road to take you there."

If you're not a competitive runner, you might dismiss goal setting as a waste of time. But trust me, goals are worth every minute you invest in them. Goals are the driver of motivation; if you have no goals for your program, you run without a sense of purpose. At some point, I guarantee it will cease to be a compelling use of time. When this happens, it becomes a lot tougher to lace up and head out the door.

Now, if you happen to be a competitive runner, goals are practically the definition of what you do on days you train—and on days you don't train, days you compete, and days you do nothing. The bare minimum of goal setting takes aim at establishing some objectives, such as improving personal best times or distances, then devising a plan (or two, or three, or four) for meeting those goals. Your entire training program, schedule, and possibly other aspects of your life such as diet and sleep revolve around helping you attain your running aspirations. Remember: You are a runner, and runners have goals.

WHY GOAL SETTING WORKS

In sport psychology, goal setting is a thoroughly researched concept. Edwin Locke, a psychology professor from the University of Maryland, is widely considered the pioneer of goal-setting research. He has performed hundreds of studies that contribute to the understanding of the science. Because of his work (as well as the work of many respected others) we know quite a bit about goal setting's impact on athletic performance. These studies suggest virtually all high-level athletes in every sport, including running, engage in some sort of goal setting, and that these same principles can inspire and guide athletes at all levels to achieve their best while getting the most from their training.

In a 1981 paper,[1] Locke suggested four main reasons why goal setting is such an important task for any athletic endeavor. First, goals direct your attention. If your hope is to, say, complete one of your running loops in under an hour, this goal will force you to laser in on aspects of your training like pace, breathing, and form. You will also probably come up with little milestones to measure how much closer you're inching to your goal, like passing the blue house with the green shutters by the 25-minute mark or noting the few seconds you shave off your time each week.

Second, goals motivate you. They give you a reason to keep on pushing forward. It's a lot easier to carry on enthusiastically and remain excited about the process if you're working toward something.

Third, it's easy to fall into the trap of viewing your efforts as a long string of failures if you're not satisfied with your performance in the moment. Having a long-term goal gives you a point somewhere in the distance to move toward. But I also recommend setting some intermediary goals you can meet during the process. Tracking the inches as well as miles allows you to taste success more frequently. This series of smaller "stepping stone" goals you meet along the way to the big payoff help transform those feelings of missing the mark into a continual string of mini successes. I would much rather you have multiple opportunities to reach success in a single run. You should, too.

Finally, having goals keeps you in the learning zone. It forces you to approach your running program in a completely different way—seeking out new resources, new equipment, and new strategies to help you accomplish your endgame. Every time you reach into your bag of tricks, you pull out new ideas you might never otherwise have dreamed up and learn new skills you weren't aware you had. This type of thinking not only develops the skills you need to reach your current set of goals, it's using and changing your brain because your brain likes new, novel experiences. This prepares you for the next set of goals, and the next, and the next . . .

As I said, goal setting is the foundation of the brain strategies to come, and the key to feeling like a successful runner. However, let me offer a small word of caution from other research: Be careful not to overemphasize goals or set them too high. This can cause anxiety and knock down your self-confidence. It's possible, even with all the ability in the world and the proper training and preparation, you will fall short. Maybe you have an off day. Maybe your wonky hamstring won't cooperate. Maybe you ate a bad shrimp the night before. Who knows? Sometimes life hands you lemons without giving you the ability to make lemonade. Seasoned runners recognize the ebb and flow of life and how it affects performance. So how you set your goals—or if you set them at all—will make you either a seasoned runner or a bucket list runner.

Resilience and learning how to manage disappointment are essential yet often overlooked aspects of goal setting. One way to factor them into the process is by having some back-up goals in your hip pocket. These are secondary goals that aren't exactly what you were aiming for—but, say, if you don't slash those seconds from your 10K time that you hoped to, you can still view weight loss or placing higher than usual in your age group as a nice consolation prize. Of course, you go into a race with those benchmarks defined as well; otherwise it's easy to discount them if you assess only primary accomplishments post-race.

One way to frame your primary and secondary goals is as best, great, and good. Your best goal is when you hit all your marks. Your great goal can be some alternate goal that, even if you didn't meet your best goal, will still give you some measure of achievement. Your good goal is one step down from great; it's something you can take away from the experience while still feeling good about your efforts. Using a race as an example, your best goal might be to set a PR (personal record), your great goal might be to run a mile within the race at a certain pace, and your good goal might be to have a strong push through a long stretch of hills or a decent kick at the finish. Make sure to avoid a line of thinking known in cognitive behavioral psychology as "discounting the positive." Simply put: Don't pull the rug from underneath any of your achievements. Claim them for what they are so you can benefit from collecting evidence that you are a runner who can achieve goals. Isn't that what you want?

However, when you train exclusively to achieve your goals you must be careful not to lose sight of everything else. Strive mightily, but don't forget to listen to your body and never stop enjoying the process. Be passionate about your mission, but remember it's okay not to live it 24/7/365.

A SYSTEM FOR GOAL SETTING

There are plenty of systems for goal setting. One you've undoubtedly seen is the SMART system, a mnemonic for specific, measureable, attainable, relevant, and trackable—or some other iteration of ideas that spell out SMART. On the whole, I think this is a useful system, but I've always felt it doesn't go deep enough into the psychology of what makes runners tick. Physiology may set the limits of your performance, but psychology determines whether you reach those limits. The body can only do what the mind can imagine.

I like to inject a bit of brain strategy into the goal-setting process, so I came up with my own goal-setting mnemonic device to reflect that. Here's how I would like you to approach goal setting:

G: Can I feel my goal in my **gut**?

O: Is my goal **objective** and measurable?

A: Is my goal challenging yet **achievable**?

L: Will my goal help me **learn** about my running and other abilities?

Let's run through this letter by letter.

G: **Gut**

What I mean by feeling a goal in your gut is making sure you have an emotional attachment to it. Work toward something you feel really passionate about achieving. In Galloway's case, he felt connected to Boston the city and Boston the race.

Your goals should mean something to you on a deep level. Whether it's a time goal, a distance goal, a weight loss goal, or something else, your goals should speak to you in a very personal way. A coach, teammate, or spouse might offer you some guidance on goals, but ultimately only you know what lives in your dreams. When you have 20,000 runners who cross the finish line of a large race, you have 20,000 different reasons for being there and for running the race the way they did.

Whatever you decide to work toward, express it in positive, measurable terms. Instead of thinking of a goal like "Don't come in last place," try something more along these lines: "I want to run my best race and finish in the top half." And when you have several goals, prioritize—try the best/great/good approach. This helps you to avoid feeling overwhelmed and gives you a greater chance of feeling success and achievement no matter what the result.

O: **Objective (and Measurable)**

Be precise and reasonable about your goals. Avoid vague, wishy-washy declarations. Building on the example above, where you expressed the goal of wanting to run your best race, a better way to state a goal like this would be to say: "I want to improve my performance in the 5K by 3 minutes by the end of the season." Including details like dates, times, and distances allows you to measure progress and record achievement both for your ultimate goal and your stepping-stone goals; hard facts and good data let you know exactly where you are in the process. And don't rely on your memory for all these numbers: Create a running journal to record your performance. By having a visual log, you help to send messages to your RAS that you are a runner.

I also recommend taking some baseline measurements. Beginning stats are important because they remove any fuzziness about progress. Don't just guess—take some actual measurements. For instance, say your goal is to run 5 miles in 40 minutes. You do a test run and find out that with your best effort you can run that distance in 45 minutes. Now it's crystal clear: In order to reach your goal, you need to shave 5 minutes off your present time. You have also given yourself a way to measure intermediary progress with precision and perhaps come up with great and awesome goals like shaving your time down to 44 minutes, or running a strong mile within your next time trial.

A: **Achievable**

Goals that are too easy to achieve don't give you much of a reason to lace up your sneakers on a rainy day. Likewise, a goal that is crazy hard and plainly beyond your reach will be discouraging rather than motivating. Research has shown this to be true almost as many times as Bill Rodgers has been asked for an autograph.

The best goals are the ones you have a realistic chance of achieving through hard work and intelligent training. If a goal takes you only a couple days to hit, it's probably not challenging enough. But if achievement perpetually seems years and years away, it can imbue you with a sense of futility; best to set your sights on some other shorter-term goals in the meantime. Choose a goal that makes you work your tail off and keeps you focused for a few months—that's the sweet spot of goal setting.

You may have to experiment before you hit upon the right balance of

WRAP-UP

I have a few additional thoughts for you about goal setting I would like you to keep in mind. As you think about goals, I prefer you focus on performance-based goals versus outcome goals. A performance goal is actually a cluster of goals that efficiently lead to an outcome goal. For example, if your outcome goal is to complete a marathon in under 4 hours, then determine the individual, smaller goals you must meet to make that happen? You might set a single outcome goal like coming in 10th place in your age group; however, you have more control over your own performance than a placement outcome. If you happen to have a bad day, or if everyone else happens to have a better day than you, even a great performance won't satisfy you.

Expect setbacks and failures. This can be especially true when you're just starting a new routine and you feel more fatigued than usual. Your times may actually backslide a bit as you increase volume or get used to new training drills. Your body will adjust, and you will gradually learn to tolerate more intensity. Be patient and keep your eye on the prize.

not too hard and not too easy. Olympic sprinter and world record holder Michael Johnson described to me his view of goal setting in a way I think hits home: "At one point, I switched from trying to win races to trying to break world records. It was a new motivation for me."

That really does sum it up. If you achieve the goal without breaking much of a sweat, make your next goal harder. If the goal took a dispiriting length of time to achieve, make the next goal a little easier. Keep tweaking. You'll get the hang of it.

Also, when determining goals you will need to take your fitness and training level into account. This is one instance where experienced runners will face more difficulty than newbies, because the fitter you are and the longer you've trained, the less room there is for improvement. When you have run for a decade or more, you have already moved much closer to your full potential compared to when you were just starting out. Even

When you reach a goal, don't just move on to the next thing. Pat yourself on the back and enjoy a moment of satisfaction as a bona fide runner. You worked hard to get from Point A to Point B, so be proud of your achievement—and if it's a really significant goal, reward yourself.

When you're ready, flip to Chapter 18 and make a few copies of the Goal-Setting Worksheet. Use this sheet to write down your goals. Putting finger to keyboard or pen to paper crystalizes objectives by giving them weight, mass, and charge. It changes them from thoughts into an action plan—"Yes, I'm really going to do this!"

Use the worksheet to conceive your running goals step by step. Post the completed sheet on your fridge, tape it to your mirror, or find some other place for it where you will see it. It will remind you daily of what you're trying to accomplish. Reevaluate as you reach your goal, or as needed. I want you to keep this in mind as well: Goals are important, but don't forget to enjoy the process along the way.

Once you've got your goals in place, it's on to Chapter 5 to learn about the next Runner's Brain strategy: visualization and focus.

with diligent training, it's unlikely you can make the kind of huge leaps people at the beginning of their careers can expect. For a seasoned runner, shaving a few seconds off a race time might be a reasonable aspiration, whereas a newcomer might aim to lower times by 10 percent or more in a season.

L: **Learn**

As Locke's studies revealed, goals that put you in constant learning mode are the most engaging and motivating. You can learn a lot about your capabilities and how hard you're able to push yourself. If you learn something that suggests you should adjust your goals, then you should do so.

Bonus: I've seen this kind of knowledge transfer to other aspects of life. I know an ultramarathoner who says her experience on the road helps her put difficult tasks in perspective. When she is given a tough project at work, she tells herself she has survived six loops around Central Park so she can survive a few extra hours of work. She says she even got through labor by comparing it to a hilly hundred-miler she raced a few years before she got pregnant!

CHAPTER 5

Visualization and Focus

From your imagination to reality

WE'VE ALREADY TALKED about goal setting as an important motivator. Now let's add another tool to your mental toolbox—visualization. This Runner's Brain strategy involves "seeing" your run in your mind's eye. You "feel" the physical exertion in your imagination and "experience" every sensation as you would during an actual run. Think of it as a dress rehearsal for the real deal.

Soviet Olympic athletes were famous for using visualization techniques as far back as the 1970s, when they dominated many sports at the world-class level. Some of their training ideas, including visualization, caught fire with the rest of the sports world; now you can find visualization embraced in the upper echelons of a variety of sports, including baseball, football, golf, swimming, boxing—and, yes, running.

American marathoner Mark Plaatjes is perhaps one of the most famous examples of a runner who used visualization as a strategy to achieve breakthrough performances. He credits his incredible (and unexpected) win at the 1993 World Championships in Germany to an aggressive training schedule and the use of mental imagery. He says he studied the course in his mind so often that even before he stepped off the plane, he knew every rise and fall, every rock, every pothole along the course so well it was like he had already run the race many times before. By the time he toed the starting line he had thought through, and planned for, every possible scenario. Because his mind was clear of distractions, he was able to focus on nothing but the run. In one of the most thrilling marathons since Pheidippides floated across the Greek countryside, Plaatjes surged ahead of the pack just a few meters before the finish. He broke the tape—something he had already pictured doing in his mind over and over.

Elites aren't the only ones who can use visualization effectively. Average runners use this technique all the time to help reach their goals. Not that simply imaging something will allow you to achieve feats beyond your capabilities. But when you can picture your goals clearly being met, you can move closer to actually achieving them. Rather than confining your progress to the moments you spend running, you can enjoy a taste of success while soaking in a bubble bath, waiting for a train, or sitting at your desk.

Still, you must run the training miles no matter what. Plaatjes has said visualization would not have worked for him if he hadn't been trained properly. As a runner yourself, you know this to be true. But my philosophy has always been that when the physical and the mental come together, you develop a performance edge.

SEEING IS BELIEVING

Scads of studies demonstrate the many contributions visual imagery can make to performance factors such as motor skills, motivation, mental toughness, and confidence. While very little research has examined how runners might specifically benefit from using visualization, much of it has enlisted some of our fellow athletes as guinea pigs.

One such study, performed at the Cleveland Clinic Foundation,[1] compared the results for beginning weight lifters who went to the gym to a group of "mental weight lifters" who only hoisted heavy poundage in their minds. The gym goers who did the real work gained about 30 percent strength, which was as expected. But to the surprise of the researchers, the virtual athletes also made gains; they increased strength by nearly 14 percent without so much as brushing past a weight. And both groups saw bigger strength improvements than a control group who never so much as gave a second thought to pumping iron.

That study didn't look at what would happen when physical and mental practices are combined—but another study[2] did. This time, New Zealand researchers recruited experienced trampolinists as subjects, classifying them into novice and experienced bouncers, then further divided these two groups into an experimental and a control group. All the athletes were then tested for their natural ability to use mental imagery as a goal-setting strategy.

Next, all the athletes trained on three bounce skills over a 6-week period. (That's one study where being a subject sounds like fun!) Each session started with 2½ minutes of physical practice, followed by 2 minutes of solving math problems, puzzles, and other games to get their minds off training, then another 2½ minutes of physical training. In the end, trampolinists who were "high imaginers" improved their skills to a significantly greater extent than the "low imaginers," regardless of whether they were novices or seasoned experts.

One more sample study[3] before we move on. This one is interesting because it evaluated groups of elite karate students to see if changing the amount of visualization might affect performance outcomes. Here, Soviet researchers divvied their subjects into four separate groups ranging from one that did all physical training and no visualization to one that dedicated about 75 percent of their training time to visualization; the other two groups fell somewhere in between.

At the end of 12 weeks, the students who did the most visualization and the least amount of actual training showed the greatest improvement in skills. In fact, all the groups who did at least some mental rehearsal beat out the group who were put through their paces with no virtual practice. I'm not sure if those results translate to a sport like running that requires fewer elaborate skills, but it's worth noting there could be a dose response with visualization, meaning the more you do it, the better results you'll get.

The point is this: These three studies—and dozens more like them—make a strong scientific case for including some visualization in your training. While there has been very little work in this area looking at runners specifically, visualization is a technique that gets universally favorable reviews in every sport where it has been tested. I feel comfortable assuming it can be of benefit to us road warriors, too.

As for why it works, that's a subject of much debate. Most experts agree that imagining stimulates the same regions and lobes of the brain that are also used in doing. Envisioning a great performance appears to create neural patterns that nearly match the neural patterns of the real physical performance. In some way this seems to code the muscle, motor, and behavioral memory of a physical skill. When combined with enough physical practice, visualization may help imprint the skills onto your lobes and help train your muscles to do exactly what you want them to do without forcing them through the actual practice. There are limits, of course. But reaching a goal in your mind's eye can move you closer to it, all while saving some extra wear and tear on your body.

INSIDE, OUTSIDE, AND BEYOND

Have you ever noticed that in some dreams you're the star and you see the action from your own point of view, while in others you're the observer watching the action as if in a movie? Visualization works the same way.

With an internal, or inward, visualization perspective, you imagine things happening in the first person. Everything you see, feel, and think happens to you. You look through your own eyes at the task at hand. With external, or outward, visualization perspective, you witness your action in the third person, as if you're sitting in the audience watching a screen or stage. A third type, known as kinesthetic visualization, combines first-person visualization with physical movements to help simulate real actions—literally moving your body as you visualize that which you want to improve or reinforce.

We won't get into kinesthetic visualization too much here. If you have ever watched Olympic gymnasts and divers on the sidelines before they compete, you have witnessed this style of visualization in action. The athlete goes off into a corner, closes her eyes, and then starts twisting and turning her body as she pictures moving through the twists and turns of her event. If internal visualization is like having the experience and external visualization is like watching a movie, then kinesthetic visualization is like playing a virtual video game. This can be quite useful for a lot of physical endeavors, but I'm not sure how much it does for a runner because running isn't a skill sport per se (although we know that is arguable on many levels). One exception to this is the kind of race that mixes in obstacle-course elements. If you're someone who runs that type of course on a regular basis, you might find a kinesthetic approach to visualization helpful.

As for the other two visualization styles, both can be productive. Internal visualization can be an opportunity to practice a running experience beforehand so you can imagine how it will feel when it's actually go time. External visualization can also be useful, in that it provides a different perspective: As the observer you can distance yourself from any negative emotions, which allows you to look outside yourself and see errors you might not otherwise spot. Isn't it always easier to see mistakes when you are watching versus doing?

I tell my runners to try different visualization practices. Switch between a few you like until you find what works best for you. You might develop a preference for one form over another, although I know some runners who toggle between different imagery scenarios depending on what they're trying to get done, their mood, and the situation.

OKAY, ENOUGH THEORY—LET'S GET TO IT

There is more than one right way to practice visualization—countless ways, in fact. And if it works for you, you're not doing it wrong, even if you don't follow a system down to the letter. The trick is to dedicate yourself to the process and make it your own. As with any skill, repetition ingrains the brain.

In fact, you will probably have to fiddle with the nuts and bolts of your practice to get comfortable. How often you use mental imagery, where you do it, and when is up to you. Personally, I have found that a practice is most effective when you do it on a daily basis for a few minutes at a time; if not daily, then at least on a regular enough schedule for it to take hold in the mind. World record holder Michael Johnson says he will visualize a single run at least 30 times before he runs it. He knows the track, the surface, the types of weather, his thoughts during the race, who he will race against, his clothes, and even hears the starting pistol. The win has happened in his brain well before anyone has shown up at the venue.

I have had some folks tell me visualization works best for them when they do it at night right before they go to sleep, others when they first wake up, and some right before a run or even during the run itself. Once you hone your powers of imagination, make visualization a part of your regular routine. For example, as you go through your warm-ups and last-minute prep, include some relaxation exercises, affirmations, and a bit of brief visualization about the run itself.

On the next few pages. you'll find five short sample visualization practices. Some are internally focused; others are externally focused. Some are a hybrid of the two techniques. Try them all. You may find that one really speaks to you or that you like switching them off. Again, whatever floats your boat. You will also find a visualization exercise in Chapter 18.

When you first begin a visualization practice, find a quiet place where you won't be distracted. Treat it as a kind of meditation—it is. It may take a few weeks to bring your imagery to life, but keep at it. You will improve. And as you become a better visualizer, you'll be able to call up your success scenarios on a packed bus, a busy street—or a crowded starting line. Remember to use all your senses in your visualization, and don't forget you get to control every aspect of the scenario. These two factors alone are what make visualization so powerful.

Visualization #1: **First-Person Success**

Begin by getting comfortable, closing your eyes, and taking deep breaths until you feel relaxed.

Now call up in your mind's eye one of the highly specific goals you established. Imagine you have just achieved this goal in the past few moments. Perhaps you just crossed the finish line of your first marathon, or you're glancing down at your watch to see your new 10K PR. Whatever that goal, hold the mental picture in your head, conjuring up as much detail as you possibly can.

Don't just see it. Feel it. Use all your senses. Is there a cheering crowd? Is someone hugging you? What are you wearing? How does your body feel? Can you taste the salt of your sweat? What are your emotions? Take a few minutes to explore all these summoned sensations in rich detail.

If you feel skeptical, you won't ever be able to reach the goal. Put any naysaying thoughts aside during this exercise. Keep doubt from creeping into the scene. Aim to see it as if it's real—and believe that it's real. Consider adding some affirmations, such as "I'm fast," "My stamina is endless," or "Nothing can stop me." Repeat this enough times and those affirmations will weave themselves into the fabric of your thinking.

At the end of your practice, open your eyes. Carry those feelings of accomplishment and positivity with you into your run.

Visualization #2: **The Success Movie**

Begin by getting comfortable, closing your eyes, and taking deep breaths until you feel relaxed.

Imagine you're sitting in a movie theater. You can even munch on some imaginary popcorn if you like. The lights dim. The movie starts. You can see you are watching a movie about you as you're in the process of completing one of the goals you have set for yourself. In this script, everything goes according to plan. You're part of the crowd watching and cheering yourself on. You see and hear everything, but from the perspective of an outsider. Notice how your body looks. Examine your facial expression. Take in every detail as if you are a spectator witnessing an actual event. Imagine the thrill of watching this scene unfold.

Now open your eyes. Think about what you just saw. Break it down as if you were giving yourself some future advice on how to achieve that same goal. Carry this information with you into your next run.

Visualization #3: **Stepping into a Movie**

Start this practice exactly the same way as the success movie. Imagine, for example, you have just watched a mental scene of yourself crossing the finish line of an upcoming race. Take in the scene as an outside observer, from your seat in the imaginary movie theater. After a few moments, picture yourself putting down your popcorn, getting up out of your chair, walking to the screen, and opening a secret door that allows you to enter the movie.

Now, experience the whole scene again from your own point of view. It's the same experience you just watched as a theatergoer, except now you're the actor. Again, see everything in vivid detail, hear the sounds you would hear, and feel the feelings you would feel.

Finally, step back out of the screen and return to your seat. Your success movie is still playing. Watch it for a few more moments, then open your eyes. Now you have a holistic perspective of your success scenario. Carry that with you into your next run.

Visualization #4: **Process Visualization**

In this visualization practice you don't project the completion of the goal onto your mind's eye. Rather, you picture the steps taken, one by one, that will take you there. A process review like this helps crystalize all the stepping stones you need to hop across to take you from where you are right now to where you want to be.

Close your eyes and get relaxed by taking some deep breaths. Now, slowly, starting at the point in your training you are right now, imagine 10 important events that must occur for you to reach your goal. For example, you might focus on several difficult upcoming training runs, a strength workout at the gym, or some active hamstring flexibility exercises you should do to avoid injury. Don't dwell on any one step too long, but imagine each step in enough detail to make it tangible. As you do this, hold on to the sense that you are attached to the outcome, i.e., your goal, so you don't lose sight of it, but honor the importance of taking the actions that will help you get there.

When you're done imaging each sub-goal as well as the ultimate goal you will reach, open your eyes. Carry this work ethic and sense of accomplishment into your next run.

A twist on this practice is to imagine running the course of an actual race, just like Plaatjes did. If possible, try to get out and actually run the course so you can make mental notes of key landmarks, potential aid station locations, major climbs, and the condition of the trail or road. If it's not possible to get out on the physical course, consider calling it up on Google Maps or a similar app; this is a good way to get a surprising amount of rich detail. Once you have committed to memory what you need to know about the course, use this information for your visualization sessions. Run through it in your imagination, picturing every step of the way going exactly as planned, right up until the point you cross the finish line.

Visualization #5: **Creating Pictures**

If you have trouble holding pictures in your mind, creating physical reminders of your aspirations can be a powerful aid. For example, you can create a vision board where you post a collage of photos and written affirmation statements of what you're trying to achieve. You might even take several pictures of yourself at the finish line of a goal race or at other parts of a course. You can use an actual bulletin board and post it in a prominent place. You might also want to try a social media site that highlights visual images like Pinterest or Instagram. If you're not an especially visual person, you can write out a series of affirmation statements on Post-its and place them in locations where you can see them throughout the day, as a reminder of your hopes and dreams.

WRAP-UP

As a runner you have a lot to benefit from visualization. Imagine how much more confidence you would have in yourself if you felt like you had already successfully finished your goal run or race a few times before. You would take a lot of pressure off yourself because a lot of your doubts would already be addressed. You would know where to put your focus and what it feels like to get things done right. You would have fewer distractions because you already know what to expect, so you would have more energy to devote to monitoring the aspects of the race that are more important.

Using imagery is also a way to deal with the unexpected during a race. You will know ahead of time what you will do if confronted with a bad stomach, blisters, a lost drop bag, or a hot day and how to react in a calm, positive manner because you have already practiced it—and it turned out great.

CHAPTER 6

Association and Dissociation

Training your brain how to respond

WHEN YOU RUN, what's running through your mind? Is the voice in your head cheering you on, or is the voice telling your feet to feel like lead? Maybe the sounds in your head are like the quiet hum of a generator. Whatever the voice is telling you, trust me: Your brain is working at least as hard as your quads, and definitely faster.

Almost everyone thinks about something when they run. Half-marathon specialist Jessica Sampson says she often gets a single thought in her head that she repeats for the entire run. "I'll start wondering something, like if I should repaint the bathroom one of these weekends, and then I'll just be thinking paint, paint, paint for miles and miles," she says.

Is she a crazy person? If so, then Lizzie Burger probably is, too.

"I count steps," Burger says. "I know I take 88 steps per minute and if I count out 88 steps 10 times then I've been running about 10 minutes."

Who knows? Maybe all runners are insane! Some runners tell me what goes on in their heads sounds like fiddling with the car radio to find a good station as they careen from thought to thought, while others say their thought process resembles the sounds of a rock concert, with them center stage singing their hearts out. Runners who choose to occupy their minds with brain waves like music, games, mantras, or conversation have layers of thought and cognitions rippling beneath the distraction. Though it may seem to you like you're having a straightforward and private conversation with yourself, those private words may have undercurrents of judgment, motivation, emotion, and memory. All these brain strategies use a pinch of goal setting, a dash of visualization, and a heaping spoonful of self-talk, as I already outlined in previous chapters. They stir up different sections of the prefrontal cortex and the cerebellum to get them whirring and humming along, all to help your feet keep moving.

TUNE IN, TUNE OUT

Just as there are different styles of visualization, there are distinct styles of thinking on the run. You can tune in or you can tune out. This is where it gets really cool and easy to learn. You can focus your thoughts inwardly on yourself or outwardly toward the world. That adds up to four distinct styles of thinking, all of them lying at your feet—so to speak.

First, with an internal association style, you focus on running performance by devoting your attention to your bodily sensations. As you're chugging along, you think about physical functions and systems such as breathing, stride, posture, and pain. You might coach yourself with some mental commands, such as "relax your stride" or "drop your shoulders," or perhaps flow through a mental head-to-toe checklist to make sure your body is aligned well and stays that way. Regardless, your mind and body are in sync as you move forward.

External associators also focus on the task at hand but place their attention on factors outside the body. When you externally associate during a race, for example, you concentrate on things like the sound of your shoes hitting the pavement or ground, seeing yourself as an unstoppable muscular hero, or thinking about the noise from the cheering crowd as fuel for your performance. All your thoughts are related to the race, just not on the physical exertion.

On the other side of the cognitive coin, runners use dissociative strategies as an escape route so they don't have to think about the run, the boredom, or the pain—to name a few unpleasantries. That doesn't necessarily mean you hate what you're doing; it just means you have a way of making the miles go quicker, by shifting your thinking to somewhere else. If you become immersed in a conversation with another runner or fired up by some great tunes streaming through your ear buds, then you're likely a dissociative runner.

Internal dissociation focuses your mind inward: You think about your problems, the past, your family, work problems, your grocery list—anything that keeps your mind busy and off running, and all that goes with it. By the way, it's possible to use visualization as a form of internal dissociation as well. Focus on the details of the Mona Lisa, design the perfect running shoe, create a website, or tease yourself with some sort of fantasy of your choosing—just keep your bib on.

External dissociative strategies distract you with mental tasks that have nothing to do with you or your run—say, listening to music, gabbing with your training partner, or counting telephone poles. When it comes to external dissociation you really have to be all in, 100 percent, to dial down whatever you aren't enjoying. Jessica's obsession with painting her room is a good example of internal dissociation, while Lizzie's step counting is a good example of external dissociation.

Most runners don't exclusively use one thinking style all the time. Without realizing it, you may switch between several thought styles during a run or use different thinking strategies in different situations. For instance, some studies show that as your pace increases, your thinking style becomes more associative, possibly because hard work forces you to monitor your body systems for potential overloads. What's even better is that you can learn to control these aspects of your brain and use them to your advantage on command when running. It's kind of like shifting gears on a bike: When going uphill you need one gear, and when cruising on a flat surface you need another.

MIND CONTROL

In general, most sport psychologists will tell you dissociation is the best antidote for boredom and association is the key to great performance. I think that's true to a point, but not necessarily as clear-cut as it seems.

Some research suggests that seasoned runners have a preference for associative-style thinking in competitive situations, for example. In 1977 researchers Bill Morgan and Michael Pollock wrote one of the first papers in sports psychology.[1] They interviewed a group of elite distance runners, asking them what they thought about in the heat of competition. They then asked the same question of nearly elite runners. Olympic runner Jeff Galloway was one of the subjects of that study.

"Bill came in and asked us a bunch of questions about what we thought about on the run," Galloway recalls. "Almost to a person, the almost-world-class athletes dissociate from the duress of the race. They had a cognitive strategy which was good, but they lived in a fantasy world so that when the going got tough, one guy was a locomotive imagining himself chugging along; another one built himself a house and distracted himself by nailing specific nails into specific places. And that worked. It got them there, but what it didn't do was allow them to tap into the intuitive part to realize their potential," Galloway says.

He goes on to recount the results from the psychological profiles of the elite runners like him: "Almost to a person, the world-class athletes associated with their experience. They were constantly in the moment. They were thinking ahead about things like what's coming up on the course, sizing up the competition and thinking about where the next water stop might be. Each had a definite strategy, which tended to work better than the ones that had a fantasy world."

Other studies of marathoners note that predominantly associative thinkers tend to post faster times and confirm that most top-place finishers in races have reported that while they did switch between associative and dissociative thoughts, they spent a far larger chunk of time tuning in.

This is not to say top runners owe all their success to their associative thinking style—though it's certainly reasonable to assume their "tune in to task" approach probably contributes in some way. Also, let's be clear: Novice runners can get some great advice from the greats but shouldn't always follow exactly in their footsteps. There is some evidence that new runners post better results using dissociation even in competitive situations.

As an interesting side note on associative-style running, I must mention Helen Cherono, a lightning-fast half and marathon specialist who is a dear friend and sometime training companion to Catherine Ndereba, one of the top marathoners in the world.

Helen is a fierce competitor in her own right who happens to be blind. In speaking with Helen, I determined she uses external associative-style thinking, albeit in a somewhat unique way.

"Especially when I am running ahead of people, I visualize how fast their feet are moving toward me. I use the sounds of the runners' feet or how hard they are breathing to visualize. Based on how they are breathing, I can tell how tired they are," she says.

I find this fascinating because it shows how, regardless of thinking style, you can use all your senses. Most runners tend to default to sight as their dominant sense, but Cherono is a good reminder that your other senses can factor in as well. You don't have to be blind to use the sound of footfalls to gauge performance. You might also use the kinesthetic sense of your feet hitting the ground or even the smell of sweat as a performance cue.

Dissociative thinking may help mitigate the tension of intensive training and competitions for people just getting used to such things. In longer races, newbies have a tendency to hit the wall sooner and may have some success using dissociation to push themselves to the finish. Once you get past the initial phase of a running program, it may make sense to shift into association at least part of the time.

As Morgan and Pollack's study indicates, most runners have a natural tendency to think in a particular way. From what average runners tell me—and even some elite runners like former Boston Marathon champions Joan Benoit Samuelson and Amby Burfoot—dissociative thoughts, at least while training, take up the bulk of their brain power. I get that. More and more we are conditioned to tune things out. We cross streets reading our e-mail, we talk on the phone in the car, we watch video at the same time we surf the Internet on a tablet. The majority of us may gravitate toward dissociation and distraction because of modern-day brain conditioning.

Of course, one of the pleasures of running is that you can let your mind wander freely, especially during training runs. This might be one of the few times your 21st-century mind isn't asked to fragment across two or three devices at a time. I never want you to lose that.

However, association may be a disciplined skill that's worth learning. You may become a better runner if you can take control of your thoughts. As I will explain in the next chapter, there are times when you want to let it flow and let your mind go. But there are certainly times when it's advantageous to take the reins of your thoughts and use them as a training and racing tool. You are very likely able to get more from your body by asking more from your brain.

PAIN AND THOUGHT

A marathoner named Sally Corskin once told me a story about running a race with something in her shoe that made each step increasingly unpleasant. Obviously the sensible thing to do would have been to stop, shake out the aggravating object, and resume running. But Corskin had her heart set on a PR. Stopping would mean losing valuable minutes and possibly killing her momentum. Not willing to blow her chances at a fast time, she decided against doing the sensible thing (stopping) and instead dialed down the pain by shifting her thinking from victim to investigator. Her thoughts changed from "Ouch, this is killing me every step" to "What the heck is beneath the big toe of my left foot?"

Over the course of many miles, Corskin ultimately assessed the item to be small, flat, and metal. She was able to determine it was round and smooth on the face but rough around the edges. By the time she crossed the finish line she had correctly guessed the offending object was a dime. (Given more time, she says, she probably could have determined the year it was minted.) Corskin got her PR—and a blister so monstrous she hobbled around for over a week before seeking medical attention.

I bring up this story because every seasoned runner I know has a similar "I toughed it out" tale to tell. Into every runner's life a little pain must fall. This we know. But pain, in my opinion, can sometimes require a special attention to thinking style. Corskin's story demonstrates the brain's limited capacity for attention, and how flooding it with external stimulus can temporarily block out discomfort or even outright pain. But her tale also demonstrates the pros and cons of dissociating pain: Is it better to pay attention to a nagging discomfort or attend to it?

It might surprise you to learn that dissociative thinking in general isn't connected to higher injury rates. It also doesn't appear to predict a history of running-related injuries in either training or competition. It could be that you can only dissociate pain to a point and once you pass a certain threshold it no longer works. You're forced to attend to what's ailing you or risk injury.

We know that runners experience various kinds of pains: blisters, side stitches, chafing, bleeding nipples, sunburn and windburn, and unexpected cramping. This goes with the territory. If dissociation helps you get through until you can finish, so be it. Just don't ignore your body's efficient ability to alert you to danger for too long. While ignoring negative feedback allows you to carry on for a while, in doing so you risk paying a price. You may not love pain, but ignore it at the peril of developing a ginormous blister—or worse.

Just as surprising, association is correlated to higher injury rates, to a certain extent. The reason for this could be that some runners who use an associative thinking style have a "no pain, no gain" mentality. Associators may expect—and even welcome—painful sensations and make the decision to gut it out.

To her credit, it appears that Corskin luckily skated just this side of the thin line of listening to her body and ignoring its warning signals. She got through her ordeal with an amusing war story and, thankfully, no permanent damage. Still, any time you push through serious pain, be it a dime in your shoe, a nagging uneasiness in your hamstring, or a growing tightness in your lower back, you might be inviting a medical bill and should pay attention to what your body is telling you.

I think it's important to learn where your own line is. Dissociation might be a great strategy for combatting the boredom of a treadmill run or for motivating you to the top of a hill. And you don't want to obsess over every little tweak of the hamstrings. But neither do you want to dissociate yourself from the throbbing pain of a calf pull for too long, or too often. Such thinking can potentially sideline you for months. There are several documented instances of runners who have completed races with broken legs.

RETRAINING YOUR THOUGHTS

The first step in harnessing your thoughts is to become more aware of them during training and competition. You'd be surprised to learn that some runners don't even know their thoughts are there. Here are three strategies for you to try.

Thought Control Strategy #1: **Notice**

To gain more control over your thoughts, try to notice thought patterns that occur in different situations in your daily, non-running routine. Start by taking note of what you think and when you think it. Then, for the next 2 weeks, you can add a column to your training journal for what you think about during your training runs and any races in that time period. There's also an app for that: You can buy one for your smartphone or check to see if your fitness tracker device has a function for recording thoughts and mood. (As these devices have become more sophisticated, many of them do.)

Thought Control Strategy #2: **Analyze**

Analyze the data—your thoughts—to see if you can spot patterns in your thinking and how they correlate with performance. If you typically dissociate, you may find it helps you through certain kinds of training runs, like long, slow distance, but doesn't seem to add much power to a hill workout. It's possible that your thoughts may be neutral, neither contributing nor detracting from your training levels. When I run steep hills, I use a combination of thinking depending on what's happening to my body. First, I'll internally associate and monitor my body as I approach the incline, then I'll switch to external association and think of my leg muscles desiring to dominate each successive stride. The closer I get to the crest of the hill, I may dissociate the burn externally by taking notice of the landscape or the view from the top of the peak.

With any type of thinking style, you also need to consider balance. Are your thoughts positive or negative? Do they support you, or do they bring you down? The way you think can have a powerful influence on your performance. If your thought process de-motivates you or creates doubts, you will have to modify it if you are going to have your brain working *for* you. If you're one those people I hear say, "Doc, you can't teach an old dog new tricks," I say bologna. Your brain is changing every minute of the day. No excuses.

Example: If you discover your associative thoughts are filled with negative self-talk, you can combat this by planning more positive alternative thoughts. That's right—just like you can change the distance, speed, and intensity of your workout, you can adjust your thoughts. You don't leave your physical preparations to chance, so why allow your thoughts to crop up in random fashion?

Here, you can come up with a list of positive statements that you repeat to yourself over and over again. You could change a thought like "I feel so slow" to "I'm working hard to get faster and it's paying off." You can work on recalling these statements every time you catch yourself thinking negatively. In this way you can actually train your brain to think more positively in a habitual way.

Thought Control Strategy #3: **Incorporate**

Whatever you find out about yourself, it will be important information for you to incorporate into your thinking style. If one particular type of thinking style doesn't appear to help you, you have three other types of thinking styles to try out. You might even try a cyclical type of thinking where you focus on your body for 10 minutes, then on external factors like the weather or terrain, then allow yourself to zone out for another 10 minutes.

Remember, you have the power to change your thoughts. You can plan and even rehearse what you are going to say to yourself beforehand, just like you might rehearse a presentation for work. The key is to stay on script, even if the situation becomes less than ideal. This is not an easy feat to pull off and will take some time to master. But you can and you will do it.

WRAP-UP

In the next chapter, I'm going to cover letting your mind go—the opposite of what we just talked about here. You may recall that your best runs were accompanied by few thoughts, a feeling of complete control, effortless movements, and a sense of being on autopilot. Sport psychologists often refer to this as a "state of flow." There's also that famous state of bliss known as the runner's high. You can't necessarily control either of these phenomena, but that's okay. Just as there's a time for thought control, there's also a time to give in to the moment.

CHAPTER 7

Higher States

Finding your running bliss

MICHELLE MEADOWS REMEMBERS her first time very clearly. She was running track for a university in upstate New York and having a tough season. She was doing her best to run hard every day, but for a period of several weeks nothing seemed to click.

Then, one mild spring weekend morning, she and a few of her fellow teammates set out for a 10-miler on the back roads off campus. It was a tough, hilly course through a vast expanse of farms and fields. Toward the second half of the run, she hit a stretch of rolling hills. That's when she got "the feeling."

"I felt light and strong and my thoughts seemed to get all happy and drifty," she recalls. "My legs felt like wheels, just rolling along without any effort at all, so happy. It was like I could go all day and nothing could stop me."

That invincible feeling lasted roughly 15 minutes, as Meadows remembers it. A car whizzed past her as she drew closer to campus and broke the spell. But, she says, she'll never forget it.

So many runners have at least one story about hitting the wall that sounds so dreadful you wonder why in the world anyone would ever want to take up running in the first place. Luckily, in running the universe balances out. For every yin, there's a yang; for every wall, there's a bliss.

Unlike the different styles of thoughts you can shape and bend to your will (see Chapter 6), some brain phenomena aren't entirely under your control. Under the right circumstances, the brain is capable of breaking out of structured thought and roaming free.

In this chapter I would like to explore two of those higher, harder-to-attain states of thought runners talk about: the runner's high and "flow." Neither of these states is necessarily spontaneous or even achievable for every runner. Some runners chase these feelings their entire careers without ever experiencing them once. "What's that?" five-time Olympian Francie Larrieu Smith asked me when we chatted about the concept of the runner's high. And Boston Marathon winner Amby Burfoot told me in his 40-plus years of running he experienced runner's high once—for eight minutes, about 30 years ago.

THIS IS YOUR BRAIN ON A NATURAL HIGH

The runner's high is a famous, though relatively uncommon, phenomenon. Since humans first began running, the occasional runner has reported experiencing a sense of peaceful euphoria wash over them during a run. Many describe it as like a drug high, but without the drugs. And in fact, the runner's high may well be a sort of drugless buzz thanks to a flood of naturally occurring chemicals that pump into your brain—and not just when you run, but during any type of physical exertion.

A class of chemicals your body produces, endocannabinoids, is related to cannabis, the active ingredient in marijuana. Scientists believe the endocannabinoid anandamide has an especially potent ability to lift mood, dull pain, and dilate the blood vessels and bronchial tubes in the lungs. When your brain and body cells release enough of these happiness molecules, you get the rush of good feelings that lead to the runner's high.

Meanwhile, the brain also gets busy releasing another class of chemicals during exercise that mimic the response to recreational drugs opium and morphine. These organically manufactured opiates, known as endorphins, have been shown to ease the aches and pains of running and activate the brain's reward systems to bring on a sense of satisfaction and accomplishment that are part of that giddy high feeling.

There could be other things happening in the body that set up the conditions for a runner's high beyond the release of natural psychedelics. For example, it appears other neurotransmitters such as adrenaline, serotonin, and dopamine may make their own contributions to the euphoria. The brain may partially shut down in response to low blood sugar; when there isn't glucose left in the bloodstream, you start feeling a little woozy and giddy, much like you do after going a long time without eating. An elevated body temperature may also somehow contribute to the happiness you feel.

Runner's high is a hot area of study at the moment, but that wasn't always the case. For a long time the scientific community was skeptical that the runner's high actually existed. They weren't willing to take a bunch of maniacs in waffled sneakers at their word, for some reason. (Go figure.) You need proof, the scientists said. But it's not as if you can hook a bunch of runners up to some sort of scan and go looking for evidence of the buzz in their brains, right?

Or can you?

It turns out that's exactly what you can do. In 2008, researchers at the University of Bonn in Germany used PET (positron emmission tomography) scans and a special chemical analysis to compare the level of endorphins in runners' brains before and after a long run.[1] Using these techniques, the German researchers were able to trace endorphin release to some specialized neurons in the prefrontal and limbic regions of the brain—the same regions that light up in response to love and other cheery emotions. The greater the endorphin surge in these brain areas, the more elated the runners reported feeling.

This study provided one of the first clear pictures of how endorphins contribute to the natural high of running. It provided some solid evidence that runners aren't merely a cult of new age hippies espousing peace, love, and sweat. This was some real science backing up the neural phenomenon associated with the runner's high.

Proof has started rolling in for the role of endocannabinoids in the runner's high, too. Not too long ago, Georgia Institute of Technology scientists put some athletes through a grueling 50-minute workout on either a bike or treadmill as they monitored the level of endocannabinoid molecules in the bloodstream.[2] They were able to pinpoint the existence of cell receptors found both in the brain and throughout the body that bind those "natural marijuana" molecules to the nervous system and set off neurological reactions that reduce pain and anxiety and create a peaceful sense of well-being.

WHY GET HIGH?

Assuming runner's high is the real deal, you have to wonder why it exists. At least, this is something I've been curious about. Pain, though unpleasant, has its advantages. It's one of the body's primary warning signals that something has gone wrong and that further movement may make things worse. It seems a little funny that the brain would have occasion to override those very important messages.

Some scientists have speculated a reason for this pain workaround. The jovial, float-y brain state of a runner's high could be left over from the time in our history when running was essential to survival. Before the human race matured into a modern society with supermarkets and home delivery, we ran to catch prey and to keep from becoming prey. The need to eat and the fear of being eaten are both pretty compelling reasons to keep up a steady jog.

Since running for extended periods of time was a necessity, the brain figured out a way to make it enjoyable, or at least tolerable. In any scenario that involved chasing a rabbit or being chased by tigers while nursing shin splints or a blistered heel, a shot of feel-good chemicals to keep you going was decidedly helpful. Some researchers looking at both animal and human models have speculated that endocannabinoids may proliferate during running to convince us how tolerable and enjoyable running can be so that we continue to do it on a regular basis.

Some research even suggests that mammals with limbs adapted for running have brains that evolved to regularly pump out endocannabinoids. It could be why we humans, as well as species like dogs, horses, and antelope, sometimes run just for the heck of it even though it comes at a higher metabolic cost and risk of injury. Animals that aren't built for running don't appear to have the same built-in brain systems. You don't often see stubby-legged ferrets go out for a jog just because.

Whatever the reason, a runner's high is like the ultimate form of dissociation and association rolled into a single brainwave; you feel no pain, yet you are one with your body. It's almost as if your brain has decided that hard physical labor is the best thing that has ever happened to it. This is incentive for you not only to keep running this time around, but also to come back again real soon for more of those magical feelings.

GO WITH THE FLOW

Runner's high doesn't discriminate against beginning or casual runners. It is picky about when it materializes, but when it does, newbies, slow-goers, and weekend warriors are equally as likely as dedicated distance pros to get high off physical exertion.

The brain phenomenon known as "flow" is a different story. It tends to visit those who have been at it a while more often than new runners.

The idea of flow was introduced by a psychologist with a tongue twister of a name, Mihaly Csikszentmihalyi (pronounced *me*-hi *cheek*-sent-me-*hi*-ee). In the early 1990s, he described how the brain can slip into a state of mindlessness such that the body can achieve optimal performance.[3] This state of being is not unique to runners or even athletes. Painters, writers, car mechanics, chefs, and anyone else engaged in an absorbing activity with which they have some mastery are all capable of getting in the zone, so to speak.

People who get into flow describe becoming so engrossed in an activity that they completely lose track of time. When they "come to" they have an incredible sense that they just experienced perfect concentration, their thoughts completely aligned with the task at hand. No distractions or mental blocks. Just a lovely sense of timelessness.

Dedicated 50-mile-a-week runner Dan Parks says he sometimes gets in the flow during early-morning trail runs near his home in Portland, Oregon. "I feel the muscles in my legs churning but it doesn't feel like work," he says. "I'm aware that I'm running but I feel light and effortless. I feel . . . fully alive, is one way to describe it."

As Parks will confirm, flow state isn't necessarily voluntary, but it doesn't exactly happen by accident either. According to Csikszentmihalyi, zoning out requires a high level of skill and experience, which means you need to put a lot of miles under your belt before your brain is primed for flow. Experience creates a familiarity with running that allows mechanical aspects like pace, effort, and stride to switch to autopilot. Experience also helps you tune out unimportant feedback and distractions. Practiced tasks take much less effort than newly learned tasks. Without the need to track every step consciously, your mind is free to roam.

For flow, you must also believe that the act of running has some intrinsic value. Running must be a deeply personal act for you. That means having a positive attitude about your running, a clear vision of your goals, and a way to measure your progress toward those goals.

When Parks gets in the zone, he's a textbook case. He has been running distance since high school cross-country (lots of experience, skill). On the days he gets there, it's usually during an early-morning run when it's quiet (fewer distractions) and he's training for an event (goal oriented, personally meaningful; a challenging yet doable workout). He finds himself most at peace on one trail in particular where he has trained for years (a route he knows by memory so he doesn't have to think about it too much).

Even with all those traits and abilities, the timing has to be just right. Csikszentmihalyi observed that flow usually happens during tasks that are challenging yet achievable. When you're in over your head, you get discouraged or injured; when you're slogging along without too much effort, you're more than likely bored. So your running pace has to be a push, but not so hard you feel like you're going to lose your lunch.

It's a balancing act for sure. Parks says it usually takes at least 15 minutes to get in the zone after he has loosened up. If some days it takes a bit of willpower to get through a run, on flow days it's the easiest thing in the world to keep going, he says. When he gets there, he usually stays there for the rest of the workout.

As Parks and others will tell you, the sense of flow is different from a runner's high. Whereas runner's high can be a feeling of joyous euphoria, flow state is more like peaceful, concentrated happiness, similar to the feeling many people have when they meditate. Thoughts are focused and you feel mindfully in the moment. Runner's high seems to be a response to novelty; flow usually a response to familiarity. If runner's high is the intersection between dissociation and association, flow is like the ultimate associative experience in which your workout becomes a singular focal point. You simply flow in the present moment with no distractions, no bad feelings, no mental blocks holding you back.

What's happening in the brain during flow state? The research in this area is slim. The little work that has been done suggests it is a condition of the prefrontal cortex, specifically in a region called the precuneus, which has been shown to be very active during moments of creativity. The precuneus is associated with self-refection and consciousness. It tends to be most active when the brain slips into autopilot and a network of systems known as "default mode" click into gear. At the same time, other executive functions in the brain somehow disengage from controlled and conscious thinking. The resulting conditions are ripe for creating the sense of concentrated free thought associated with flow.

There is also some discussion in the research about the certain type of brainwave that is present during flow. Your brain has four main types of waves—alpha, beta, theta, and delta—that operate at different speeds or cycles. When you are wide awake, alert, and thinking at a conscious level, your brain hums along at the beta level at about 13 to 25 cycles per second, the research shows. During activities like meditation and relaxation, the brain slips into a more laid-back 8 to 12 cycles per second. This is on the alpha spectrum, and is the one in which people have traditionally reported feeling a heightened sense of concentration and creativity. It's the wave most associated with alertness, imagination, and sharp memory. And it's the wave in testing that appears to be present when someone is in the flow state.

I can't go any further without telling you quickly about the easiest breathing exercise for relaxation that I know. If you would like to try breathing to relax, the easiest way to learn is four-square breathing.

Picture a square in the air in front of you. You are going to use the sides of the square to guide your breathing. When you go down the right side of the square, you exhale to the count of four. When you go across the bottom of the square to the count of four, you neither inhale nor exhale, you just "hold it." As you go up the left side of the square in the air, you inhale as you count to four. Then across the top of the imaginary square, you hold it again until the count of four. Visualize how you are pacing your breathing and using an object to do so. I often use that very technique before doing a live radio or television interview. I also do it backstage before speaking to an audience. You, too, can find ways other than running that four-square breathing can help you. Remember, when you inhale, you want to fill your stomach, not your chest, with air.

MAKING RUNNER'S HIGH HAPPEN

You don't necessarily have control over the runner's high. Some people who love running never experience it. But you can create certain conditions that are favorable to it. Even if you don't get all the way there, the following strategies are good habits to inject into your conscious and deliberate thoughts.

Runner's High Strategy #1: **Think Positive**

If you bring any kind of negativity with you into your workout, it's unlikely you will be able to let yourself go and feel a natural high from running. Filling your mind with constructive thoughts can prime the pump of happy brain chemicals.

Runner's High Strategy #2: **Look for Novelty**

If you run the same course, at the same pace, at the same time every day, it's going to be tough to beat boredom. One of the best ways to get the brain popping is by mixing it up. Experiment with a longer distance, a fartlek (combination of fast and slow running) workout, sprints on the track, or hill repeats. You may still wind up staying with your prefrontal conscious thoughts, be they associative or dissociative, but your run will be more interesting.

The reticular activating system (RAS) constantly seeks out information that is new and different. When this loose network of neurons that spreads throughout the brain latches on to something novel, it helps block out distractions and focus your attention, setting up more ideal conditions for if not a runner's high, then certainly a more engaging experience.

Runner's High Strategy #3: **Go Hard**

There isn't any magical distance or pace for a runner's high, but most runners say it comes to them when they work out harder than usual, like during a challenging long-distance run, a shorter, high-intensity interval session, or toward the end of a hard-fought race. Evidence suggests that a taxing workout is the most likely way to experience a runner's high.

Endocannabinoid production increases when you're a little bit stressed but not pushed to the brink. Running at 70 to 85 percent of your maximum heart rate seems optimal for producing cannabinoids, as well as the so-called fight-or-flight hormones such as cortisol and adrenaline. Endorphins, too, have a higher production at that level, in response to the pain of exertion.

Your runs shouldn't be excruciating, but you should be pushing to the edges of your comfort zone. You also shouldn't be pushing hard every day to the point of chronic stress and an overproduction of hormones.

Runner's High Strategy #4: **Run with Friends**

Hook up with some running buddies (we'll discuss this more in Chapter 10). An Oxford University study found that rowers who exercised together bumped up endorphin release significantly compared with solitary rowers.[4] Meadows says her feelings of euphoria almost always come on while running with a group. However, on your own, consider wearing headphones and cranking some of your favorite tunes. Research shows that listening to music you love is another good way to elevate endorphin levels. Research also suggests that endocannabinoid levels are three times greater first thing in the morning compared to the evening, which suggests mornings are a better time to find your bliss than later in the day.

TUNE IN TO ZONE OUT

Does it seem odd that you have to work hard at zoning out? Like any useful skill, flow can take time to master. If you don't have much experience flexing your mental muscles, it can take time to whip them into shape. In that way, the brain is really no different than your quads or your glutes. You will need regular practice and training for any chance of achieving flow. The more you train your mind to zone out, the better chance you have of getting there.

To give yourself a chance of feeling flow, you will have to do some of the same things you do to encourage runner's high, but also a few things differently. You still need to go into your run with a great attitude. You still need to check your negative baggage at the door with your street clothes. You still need to think positive thoughts, let go of your anxieties, and stay in the moment. But you also need to bring with you discipline, experience, and focus. And again, there are no guarantees.

Flow Strategy #1: **Associate**

Runners tell me that flow is more apt to grace you with its presence when you begin your run with a conscious, controlled stream of thought. Try to make your thoughts associative rather than dissociative so you are really concentrating on the task at hand. Minimize distractions and multitasking. Leave your worries, problems, and phone at home. Although flow has a sense of timelessness, it also has a sense of complete submersion in what you are doing.

Flow Strategy #2: **Strive**

According to Csikszentmihalyi, stretching slightly to reach a goal that's just a bit more advanced than where you are now is a good way to cultivate flow state. So you should try to find the perfect challenge. If you typically run 5 miles at a consistent basis, trimming a minute off your time or aiming for 8 miles are aspirations you will have to work a bit to achieve but that should totally be within your abilities.

It also helps to have clear goals. Saying you want to get faster is one thing. Saying you would like to shave 30 seconds off your best 10K time or complete a marathon in under 4 hours is specific, actionable, and measurable. The goal should be something you really care about. It should be fun. And it should be truly achievable—not easy, but with some hard work and elbow grease you should be able to get there. Not only that, you also have to let go of any preconceived limitations and *believe* they are achievable. (See Chapter 4 for more on why goals are important and how to set them.)

However, as you run, stay in the moment by focusing on the process of running rather than the objectives you've set out. These may seem like a contradiction, but it isn't. The easiest way to explain it is that you should try to enjoy the journey even as you deliberately head toward your destination.

Flow Strategy #3: **Choose Your Company**

Some people say they have an easier time getting in the zone on solitary runs. Others tell me they feed off group energy. I think you will have to experiment with that. While running with the pack is distracting for some, others draw strength and motivation from it.

WRAP-UP

It's easy to get hooked on the idea of these two higher brain states. Once you have experienced the freedom of either feeling, you'll want to return to it again and again. During your runs, your brain will be filled with deliberate and ingrained thought patterns most of the time. Learning how to spark joyous and energetic or happy and relaxed brain reactions isn't always possible, but they are definitely worth pursuing. And even if you never get there, it's worth trying. The work you do will pay off in making you a better runner. At the very least, your runs will be more productive and enjoyable.

CHAPTER 8

Lucky Underwear and Enchanted Shoelaces

The power of magical thinking

NOT LONG AGO, I posted a question on the *Runner's World* Facebook page asking runners if they have a ritual, superstition, or silly belief they engage in before a race. In less than 3 hours, I received 187 replies.

Forty runners said they have to eat the same exact thing before a race. For four of them it's a pre-race banana; for seven, it's a bowl of oatmeal; three said they always eat salmon and broccoli the night before; and one couldn't line up at the start without slurping down a bowl of Cinnamon Life cereal.

Twenty-three people admitted they always, always have to go to the bathroom before a race. A couple people have to go more than once—one person confessed to peeing exactly six times before the gun goes off.

Sixteen runners said they felt the need to wear a certain type of makeup or jewelry (not all women, by the way), while several said they don't feel right unless they are wearing a special ponytail holder, hairband, bandana, or hat. Five men said they always shave their face; five women said they always shave their legs. Five have lucky safety pins for their race bib, and another four feel compelled to write a lucky number somewhere on their body or shoes.

Fifteen soulful racers pray just before they line up for the start. Eleven are very particular about hearing a specific song or playlist. One admitted he can't have a good run unless his wife takes his picture in the exact same clothes standing in the exact same pose just before the start.

Twelve runners said they consider it bad luck to wear the race T-shirt before they actually finish the race. One said if you don't finish the race, it's considered bad luck to ever wear the T-shirt. Another has to wear a Star Trek tee for anything over a 10K. Five people sport a special pair of lucky underwear for every race they run.

Seven people have a pair of lucky mismatched race socks. Eight people say they can't run a race without retying their shoelaces, often multiple times. For luck, eight people said they always put on their left shoe first. Five said they put on their right shoe first.

There were mentions of all manner of lucky pennies, magical colors, a particular way to wear the race bib, and a compulsive need to touch trees, street signs, and even other runners at a certain mile marker along the course.

These average, everyday runners aren't the only ones with their lucky charms. Francie Larrieu Smith said she had a thing for a while where she needed to wear white for a competition, and in her twenties she went through a period where she had to wear a certain pair of earrings. Joan Benoit Samuelson had her famous lucky painter's cap. Even Meb Keflezighi admits to having a pair of lucky socks and a lucky meal.

Are you getting the idea that runners are a superstitious lot? They are! (In general, athletes of any sport, as creatures of repetition, tend to be superstitious.) It seems some have more rituals just to get out the door the morning of a 5K than a pilot has items on a checklist before takeoff. And as long as we're being honest here, let's fess up: How many of you believe that not tapping something at the turnaround point of an out-and-back course means the mileage doesn't count?

BIRDS OF A FEATHER

Noted psychologist B. F. Skinner began one of his famous lectures[1] by covering several pigeons in cages equipped with automatic feeders that delivered food every 15 seconds. When he uncovered the cages, one bird would turn counterclockwise three times before looking into the food basket. Another would thrust its head into the top left corner. Another would dip its head and lift a foot. Every time their cages were uncovered and they were given access to the food, each pigeon danced through its own unique ritual.

Skinner had an interesting explanation for this strange behavior: Although we know the food is delivered regardless of what the pigeon does, the bird doesn't. The poor bird is plagued with anxiety over where its next meal is coming from, so it creates a ritual dance with movements it has done in the past right before the food appeared as a sort of good luck charm—an avian superstition, if you will.

Feeling superior to those poor, dumb birds right about now? You shouldn't. Skinner postulated that a pigeon's pellet dance is not unlike the rituals and superstitions we humans create to calm our own anxieties. And yes, I'm looking at you, person, who is compelled to eat a bowl of Cinnamon Life cereal right before a race. You, too, Bostonians. If big brother was raiding the fridge when the Red Sox scored, now big brother has got to stay in the kitchen every time the Sox have someone on base.

Regardless of species, it seems superstitions become behaviors when the brain repeats whatever actions preceded past successes. This is true even if the repeater cannot fathom how these actions could possibly have an influence over the outcome of an event. A runner who believes wearing a certain pair of underwear is vitally important to his performance, or that touching a stop sign at the 9-mile marker makes a difference in his race time isn't all that different from the pigeon hopping on one foot in hopes of a snack.

IT REALLY IS ALL IN YOUR HEAD

In the most basic sense, superstitions are a type of brain habit. Your brain clings to a habitual behavior because it would prefer not to find out what happens if you let go of it. In your mind, the ritual benefits you either by ensuring something good happens or by helping to avoid something bad from taking place. When you perceive the stakes in a situation are especially high, like the outcome of a race, there is even more pressure on the brain to repeat whatever behaviors preceded a good result in the past or must have prevented something from going wrong.

Some scientists believe this sort of magical thinking has a neurobiological basis.[2] The reticular activating system, or RAS, seems to be the part of the brain that helps you sort information from which you derive beliefs about yourself. It provides you with the rationale for how you see yourself, and part of that image includes your beliefs, whether they are based on reality or superstitious thinking. Some of the superstitious beliefs about clothing may be due to a phenomenon known as enclothed cognition; that is, you develop beliefs that associate how you feel and your self-image with certain articles of clothing.

Parts of the frontal lobe also seem to be wired for observing purposeful action and decoding the emotional impact of those observations. In particular, the caudate nucleus, a tiny structure situated near the center of the brain, is packed with dopamine neurons designed to fill your mind with happy feelings and other brain chemicals that influence superstitious behavior.

Dopamine, you may recall from previous chapters, is the chemical that floods your brain when you experience something that makes you feel good. But dopamine also helps coordinate emotional response with movement so you not only build the emotional connection to perceived rewards, you also take action to move toward them. In one well-known Swiss study, scientists found that high levels of dopamine were associated with paranormal thoughts and a strengthened ability to detect patterns.[3] And as an important chemical in the brain's reward and motivational systems, it helps zero in on information the brain considers relevant to a situation. Paranormal thoughts, repeating patterns, and relevance are the building blocks for superstitious beliefs, particularly if your bib number is 666 or ends in a 13.

Scientists also suspect a genetic basis for superstition in the form of a gene by the name of VMAT2. The gene seems to be the switch that regulates the flow of serotonin, adrenaline, norepinephrine, and dopamine—the brain chemicals that are the usual culprits in an emotional response. Interestingly, these same chemicals play a role in athletic performance as well. For example, adrenaline, the so-called flight-or-fight hormone, elevates physical response; serotonin is a mood enhancer; dopamine and norepinephrine are involved with mood and behavior.

Some preliminary research shows that people who have a particular variation of the VMAT2 gene are predisposed to releasing greater levels of these feel-and-do chemicals. For someone with this genetic variation, performing some type of ritualistic behavior would open the floodgate of brain chemicals, ostensibly increasing the chances for a better physical performance. Someone like this would have a lot invested in keeping the superstition going.

This area deserves more study, but right now it's at least an intriguing theory.

THE LUCK OF IT ALL

Runners certainly don't own superstitions. Rituals, good luck charms, and jinxes are found in all cultures. This includes the subcultures of various sports, which are riddled with playoff beards, pregame meals, and special numbers. Some famous examples: Tiger Woods only wears red shirts during final rounds. Serena Williams always bounces the ball five times before the first serve and twice before the second. Hall of Fame baseball player Wade Boggs is really superstitious—he had to eat chicken before every game, took precisely 150 practice hits before each game, and always wrote "chai," the Hebrew word for life, in the dirt before each at-bat.

We don't think about whether they are rational, or even if they really have an effect; we just continue knocking on wood or throwing salt over our shoulder or some other equivalent because in the deep recesses of the brain, we have linked these rituals to success or the avoidance of failure. So why is magical thinking so prevalent in athletics?

To some extent, it appears to work.

Research seems to support the idea that going through absurdly pointless motions can have a positive effect on performance. One *Journal of Experimental Psychology* investigation[4] asked a group of college students to "jinx" themselves by saying out loud that they definitely wouldn't get into a car accident this winter. When asked about it at a later point, the jinxers thought it was more likely they would get into an accident compared to the kids who didn't jinx themselves, probably because the jinx brought the idea of a crash to the front of their minds.

The researchers then asked some of the students to cancel out the jinx by knocking on a wooden table as a way of inviting in good luck. In this instance, the table knockers were no more likely to think about getting into a fender bender than those who hadn't jinxed themselves in the first place. This time-honored superstitious behavior reversed all the bad effects of the jinx, at least in the students' minds.

The researchers suspect that knocking on wood and other rituals share a common purpose. Underwear, shoelaces, and cereal may not really possess magical powers, but in your mind you believe they will help "push away" bad luck. The researchers suspect that people tend to develop superstitious thoughts and behaviors to help calm the mind. These seem to stimulate the feelings, thoughts, and sensations you experience when you want to avoid something bad from happening.

That's not really magic, by the way—as I said, it's partly something that is hardwired into the brain. But in my opinion, it has a lot to do with beliefs and emotions, too. We tend to give meaning to objects and actions even if they aren't completely rational. Once you make up your mind about such things, it's hard to be convinced otherwise. Circumstances conspire to confirm our beliefs over and over again because we all have a sort of confirmation bias—that is, we tend to pay the most attention to the instances that confirm our beliefs. You might not get a PR every time you tie your left shoelace first, but if every time you ran a good race you happened to tie your left shoelace first, you'd be hooked.

This type of thinking, illogical though it may be, offers a sense of control, comfort, and meaning, not unlike the way a child feels when he clings to his favorite teddy bear. Instead of allowing your anxiety to get the best of you and possibly hamper your performance, the enchanted beliefs and ritualistic behaviors of a superstition help keep you calm and focused. This in turn actually does affect your performance in a positive, if somewhat indirect, way.

At the University of Cologne in Germany, researchers devised a series of experiments to demonstrate the power of this kind of magical thinking.[5] In the first test, the subjects that were given a "lucky" golf ball sank 35 percent more putts compared to the golfers who played with ordinary golf balls. Then, in a second test, the participants were tasked with putting as many tiny balls through the holes in a cube as quickly as possible. Those who were told to cross their fingers for luck did better than the ones who just went for it. In the final two experiments, subjects who were told to bring their own personal lucky charm to the lab with them and were then put through a series of memory tests performed better than the subjects who didn't.

As the study shows, superstitious beliefs reinforce an athlete's confidence in his or her own abilities. Also, they add order and meaning into a chaotic and absurd universe. In the athlete's mind, a pre-competition routine is clearly something that should lead to a positive performance outcome. I couldn't agree more.

That doesn't mean superstitions have no downside, however. If you build up their importance too much in your mind, they can cripple you with distraction and intense anxiety. You have been defeated before you ever compete. You should be the one to control your special talisman—

it should not control you. What happens if you can't find your lucky underwear? Or if you get assigned a bib with an unlucky number? Will that undermine your confidence and hamper performance, or freak you out so much you can't even bring yourself to race?

At its worst, superstition is a form of mental hoarding that can lead to obsession, fatalism, or abject terror. But on the other hand, maybe the angst of living without them would leave you paralyzed with fear every time you toed the starting line. I think you can run along the thin line here. Have your superstition, but be flexible. Keflezighi, for example, told me that early on in his career he always had to eat spaghetti and meatballs the night before a big race, and it always had to be the same type of pasta. But as his career took off he began traveling around the world.

"You can't always get pasta in Japan," he says. "I still prefer to eat spaghetti and meatballs because it is a tradition, but I've learned to be okay if I can't get it. I don't let it affect my race."

So, yes, I think it's okay to believe in "magic" to the degree it reminds you of actual abilities you possess as a runner. It doesn't make you ignorant or irrational—or even a pigeon. But it might help you run a good race.

WRAP-UP

Most everyone is prone to magical thinking, and there's nothing wrong with it as long as you use it to your advantage. If you believe pinning your race number on a certain way or wearing a certain pair of socks will get you a PR—go for it. You're in good company with the rest of your fellow runners, including those at the world-class level. Just don't freak out if you can't find your special safety pins or if one of your lucky socks has gone to dryer heaven. So long as you don't let your superstitions sidetrack your thinking too much, they may help you stay relaxed and focused.

CHAPTER 9

An Uncommon Thread

How your running clothes can exert a psychological influence over the way you run

"SUIT UP!" That's Barney Stinson's catchphrase on the hit TV sitcom *How I Met Your Mother.*

Stinson (played by Neil Patrick Harris) was so dedicated to suits he sang songs about them, wore suitjamas to bed, and once described his clothing of choice as "full of joy . . . the sartorial equivalent of a baby's smile." From the first time he smoothed his tie and shot his cuffs, Stinson was defined by what he wore. If clothes make the man, he was all gabardine wool and pressed silk.

Stinson may have been a fictional character, but the power clothing is very real. It's a bona fide, documented effect that what you wear has the power to change you and, by extension, your workout. This is a relatively new area of study, and it is highly technical and theoretical. I would like to dive into the science here for a bit to make a case for why you should choose what you wear to run in with care. This goes beyond the superstitions of having a pair of lucky underwear or magical socks. It's closer to the idea of building a runner's identity.

A SERIES OF EXPERIMENTS

Scientist Adam D. Galinsky, a professor at the Kellogg School of Management at Northwestern University, was the first to identify the phenomenon he dubbed "enclothed cognition" in a series of clever experiments done in 2012.[1]

In the first study, Galinsky's lab explored what the effect of wearing a white lab coat had on attention span. He divided 58 undergraduates into two groups. One group was asked to wear a lab coat and the other was allowed to stay in their regular street clothes. After the students in lab coats were informed that their crisp, white garments belonged to a physician, both groups were tested on their levels of "selective attention."

To do this, the scientists scored the student's ability to spot inconsistencies in a series of tasks—for instance, noticing the illogic of the word "red" written in green ink. The subjects who wore the pristine white lab coats made half as many errors as those who wore their everyday clothes.

A second experiment randomly assigned another set of students into three groups. One group was asked to wear a doctor's lab coat, another was given the exact same item of clothing but told it belonged to a painter, and the last group wore regular clothes but was shown a doctor's coat draped across a desk. Once again, all three groups were given a series of tests to measure their sustained attention—and once again, the subjects wearing the doctor's coat scored the highest.

Finally, in a third experiment, students were asked to write down their thoughts about a lab coat while wearing either a doctor's coat or a painter's coat, or while wearing ordinary clothes and contemplating the doctor's coat draped across a chair for several minutes. As in the previous two experiments, those who wore the coat they believed belonged to a physician demonstrated superior sustained attention skills.

Interesting. But What Does It Mean?

Galinsky concluded that certain items of clothing carry symbolic meaning. While painting is a noble profession to be sure, most people in Western societies don't associate it as closely with reasoning, attention to detail, and focus as they do with physicians. When you wear clothing that is typically worn by a doctor, you subconsciously take on the traits traditionally attributed to a doctor. You have to wear the coat to have its strongest vibes rub off on you, the scientists concluded. Simply looking at it isn't enough to get the same effect.

Whether you know it or not, you have seen or experienced examples of enclothed cognition in everyday real life. The symbolic and social nature of clothing has surely exerted its influence on you in an untold number of ways. To wit: Recent research done at Northern Illinois University found that negative moods have a greater sway on what a person wears than positive moods, at least for women.[2] One study found that depressed women were more likely to throw on a pair of jeans and a baggy top in order to try and "hide" themselves, whereas women with high self-esteem were more likely to wear flashier clothes as a reflection of their feel-good attitude. This suggests that women, at least, have happy wardrobes and sad wardrobes, and that depressed women put a lot less effort into their appearance than women who feel happy.

Dressing for a job interview is another great example of how clothing can help you inhabit a role. Most people know enough to dress in a businesslike outfit when applying for a job, even if in their normal life they walk around poured into a skin-tight leather jacket with a ring through their nose. While dressing in a neat and conservative suit broadcasts to the world (and your prospective employer) what a skilled and successful professional you are, that same outfit can also help you reflect those same attributes back on to yourself. If you have ever been in that situation, you probably have experienced how buttoning up your suit jacket and slipping on a pair of spit-polished shoes suddenly transforms the way you carry yourself. Hey, that Stinson was on to something!

Clothing color also has the power to change perception. Different colors appear to have different effects. Interestingly, studies show that teams who wear red uniforms are likely to dominate teams who wear blue uniforms. In this culture we associate the color red with high levels of testosterone and dominance, so perhaps it helps someone wearing red to feel more dominant, and his opponent to feel weaker.

Of course, enclothed cognition aside, others will judge you on what you wear. We all know this. Studies show that children as young as seven believe clothing choice communicates something about the person wearing it. One Turkish study found it takes just 3 seconds for people to make a snap judgment about others, and much of that judgment is based on what someone is wearing.[3] Minor differences, such as a slight color change or a tweak to cut or sizing, can shift these judgments hugely. Even teachers have been found to judge a child's intelligence on how attractive they look.

HOW YOU CAN USE ALL THIS INFO TO YOUR ADVANTAGE

Perhaps it's obvious to you how enclothed cognition might work for a runner. Dress like a doctor, you'll pay more attention; dress like a runner, you'll run better.

Slipping on a sleek pair of tights does more than just make your butt look shapely. It makes you feel like a runner. Everyone instantly identifies you as a runner—and you identify yourself in that way, too. You have just fed some valuable information to your reticular activating system, or RAS, the part of your brain responsible for forming beliefs and contributing to personal identity. Just by virtue of what you wear, you send messages to your brain that you feel more inclined to go a little farther and push a little harder. The right clothes prime the RAS to help you believe you are faster and stronger, more like an athlete.

With this observation in mind, you might be able to improve your running simply by changing the way you dress. You dress the part, you feel the part, you are the part.

Think about what kind of symbolic meaning each article in your running stash holds for you. Now ask yourself: *What makes me feel fierce, fast, strong, and confident?* Does one especially sleek and streamlined pair of shoes make you think of speed and stamina? Does your ultra-techy parka remind you of hiking in the mountains and thus inspire you to get out and run hills? What about those lucky socks we talked about in Chapter 8? Does wearing them give you a feeling of invincibility?

Perhaps allowing superstition to guide some of your clothing choices is a bit of voodoo and magical thinking, but it makes sense to choose the article of clothing that you most closely identify with success in your running.

If you want to get really wonky about it, try mindfully incorporating the findings from the enclothed cognition experiments to intentionally shape your subjective psychological experience of getting dressed for a run. As you are getting ready, you could take a moment to check in with yourself and ask: *What do I want to feel like today?* With that thought in mind, and with some mindfulness about your clothing choices, you're halfway there.

Once you have identified the article of clothing that symbolizes what you hope for in your run, put it on. Accessorize with the complimentary hat, sunglasses, and even water bottle that most accurately match your desired state of being. The more confident you feel, the more apt you are to get out on the road, climb on the treadmill, or head to the track and work on getting the results you're looking for. You will start feeling better, and that in turn will inspire you to dress the way you feel. Your clothes represent your inner motivation and feelings. It's a feedback loop reinforced by the RAS and memory areas of the brain: *I feel good, so I'm going to wear the things that make me look good.*

Try this, and notice how your intentional enclothed cognition experiment evolves. Do you see any behavioral changes? Or is getting dressed simply more fun?

This is not to say enclothed cognition is the same as motivation. It's unlikely a new pair of pants will inspire you to get out onto the track if you haven't been able to drag yourself there in the past 6 months. But it might.

WRAP-UP

You'll be amazed at how much enclothed cognition can influence your view of yourself as a true runner. The right outfit can make you truly "feel" like you belong. As in any other situation in life, when you dress the part, you become the part. That's why you should choose outfits that make you feel like you belong out there on the road.

PART 3

TRAINING AND RACING

SOONER OR LATER, every runner flirts with the idea of kicking it up a notch. Will running with a partner push you to become a better runner? Should you test out your training by entering a race or two? And if you decide to race, what does it take to tackle some of the great courses every runner dreams about? These are the questions I have taken on in this section. I'll help you answer these questions in a way that makes the most sense for you. And I'll offer plenty of food for thought.

CHAPTER 10

Pack Mentality

Training with partners and groups

THE LONELINESS OF the long-distance runner is one of running's most ingrained clichés. Of course, as a runner you know this is not a saying that comes out of left field. In a recent *Runner's World* survey, 73 percent of readers reported running alone most of the time. Just 7 percent said they ran with a friend on a regular basis, and a mere 5 percent said they usually ran with some sort of group or club.

I get it. In a world full of chirping cellphones, bleeping text messages, and the constant chatter of friends, family, and coworkers, the quiet echo of your footsteps can be a welcome respite. Not that spending time with friends or friending acquaintances on Facebook isn't great. But sometimes you get the urge to unplug and go it alone with your thoughts. For many of us, hitting the road means hitting the mute button on life. It can be a rejuvenating experience.

I also have a theory that runners aren't team players. Don't get me wrong—most runners I know are the loveliest people. It's just that they don't seem to get a thrill from passing a ball, strategizing a play, or lobbing one over the net. They would prefer to leave the maneuvering, the coordinating, the compromising (not to mention the equipment) out of their workout. To them, a solitary run is "me, myself, and I" time, a personal reboot that affords them a bit of breathing room from the more collaborative aspects of their lives.

Along with this lone-wolf tendency, so many runners I know are goal-oriented creatures, too. You may not see it right away, but those two pieces of information are related. Goals, as I've told you, are the foundation of running psychology. Whether you run solo or social can make a difference in how you reach those goals.

THE SOCIAL EFFECT

As John Davis approached his 40th birthday, he began thinking about doing something big to celebrate the milestone. He had done a couple marathons over the 7 years or so since he had seriously taken up running, but he never really focused on his race times or his training pace. He thought he deserved a pat on the back just for being out there on a regular basis, he tells me.

"With my big four-oh looming, I decided to get a little hardcore," he says. "Marathons I had run up until then were somewhere in the four-and-a-half-hour ballpark, which was fine. But when I started thinking about what I could do for my birthday, I wondered if I had it in me to break 4 hours."

Davis occasionally ran with some buddies on the weekend, but for the most part he did his own thing in the early morning before work. It was convenient and uncomplicated, he says, though he admits he could sometimes be lazy about thinking outside the box. He pretty much ran the same route at the same time every day and rarely pushed the pace.

"For a long time I was going on autopilot," he says.

With 6 months to go before his birthday and this goal of breaking the 4-hour barrier, he needed some motivation and decided that joining the local running group that met in the park near his house a couple nights a week would do the trick. This meant changing his routine, but he figured he could swing it, especially if there would be a payoff. Anyway, what could it hurt to join a few workouts, he thought? If it didn't work out, he could always go back to what he was doing.

Psychologists will tell you John had the right idea. They have known for decades that exercising with a partner or a group can have a potential positive effect on performance. As far back as 1898, psychologist Norman Triplett noticed that cyclists who raced in pairs rather than against the clock increased their speeds, and children who competed against each other to wind up fishing wire on a reel did it faster than when they were left alone to do the task.[1] (Too bad his name wasn't Doublett—how perfect would that have been?)

Triplett called this partner phenomenon the "co-action effect." Psychologists soon discovered that simply having an audience can be enough to increase performance, due to a related psychological response known as the audience effect. Both effects are a form of social facilitation defined by improved performance due to the mere presence of others.

Social facilitation, the research shows, is often a powerful motivator. In one study,[2] cyclists were told their heart rates were going to be measured based on their performances. One group of cyclists pedaled alone on a stationary bike while a second group was connected with a partner via video chat. A third group was also partnered virtually, but they were told their performance would be judged on the partner who turned in the weaker performance.

Before the cyclists hit the road (metaphorically speaking), researchers added a twist to the experiment by telling some of the subjects a little white lie. Though they believed their companion could also see them projected on a screen and would be working with them to race against the clock, in reality their virtual buddy was prerecorded. Cyclists in the third group, the one where subjects thought they were being judged on performance, were paired with a looped recording so that no matter how hard they cranked it out, their "partner" was always the better cyclist.

This is what happened: The solo cyclists pedaled mightily before conking out just before the 11-minute mark. Pretty good. The second group, who felt pushed by their prerecorded partner, lasted just over 19 minutes—a not-too-shabby 87 percent improvement over the soloists. And the third group? They tried in vain to keep up with their superior (though fake) companion. They pedaled their hearts out for just under 22 minutes—twice as long as the lone cyclists.

Assuming these social influences translate for runners—and I think they do—imagine having someone there day after day to push twice as hard as usual. It's a safe bet you would see pretty massive gains in your performance before too long.

The more the merrier may also be true. In follow-up studies, the researchers showed that exercising in a larger group could potentially lead to even bigger improvements than pairing off with a single partner. Multiple studies elsewhere have found that running with a team or running group on a regular basis is motivating and improves performance among athletes. There is also some evidence that experiencing a runner's high is more likely to happen when you run with a partner or group, especially if having company pushes you to run somewhat harder than usual.

Furthermore, Michigan State University studies show the same social effects hold up whether your workout partner is alongside you or projected on a screen. This effect is especially strong among people who belong to the same team. The Michigan State studies[3] found the slowest members on the team often made the biggest gains in performance during competitions, coming through when their team needed them the most.

Hearing all the positives about social running, you might conclude that it's time to abandon your lonesome ways and get yourself a wingman, or perhaps an entire squadron. But the social effects of running aren't necessarily advantageous in every circumstance.

For one thing, it matters who you run with. When Santa Clara University researchers had test subjects run or bike next to a girl who was all dolled up with makeup, jewelry, and a fancy running outfit, they ran slower than when they ran next to the same girl sans makeup and wearing baggy attire. Subjects reported feeling slightly intimidated by the polished runner and backed off their effort because she "looked fast." Likewise, runners feel motivated to run harder with runners who are slightly faster than they are but tend to be put off by a partner who can easily cream their corn.

Davis, for example, found that he did indeed work harder on the days he met up with the running group. He stretched himself to keep up with a slightly faster subgroup that paced him a little outside his comfort zone. At first he lagged slightly behind the main pack, especially during the sprints and pickups, but gradually over a few weeks he was able to build up enough strength and stamina to hold his own somewhere in the middle. As a bonus, along the way he picked up a bunch of tricks and tips from other runners that proved valuable in both training and racing.

Thanks to the boost in his training, he broke the 4-hour mark with 8 minutes to spare, about 3 weeks before his birthday. Davis was thrilled with this outcome, of course (especially for an old guy, he says), but admits that running with a team wasn't all sunshine and unicorns.

"I found it kind of a pain to change my schedule in a way that wasn't necessarily the most convenient, and the workouts took up a lot more time because you're waiting for other people to show up and then you're gabbing afterwards . . . it wound up being kind of a time suck," he says.

And while doing the team workouts twice weekly made him a better runner, Davis says he also had a lot more aches and pains than normal. When trying to keep up with a slightly faster crowd, he sometimes let his mind wander to other things to the point of ignoring some of his body's red flags. Within a couple of weeks, he says, he started to feel slightly burnt out and found that he was perpetually aware of his Achilles tendon, which, while not exactly injured, throbbed constantly.

In addition, his goals didn't always match up with the group's. Some evenings he would show up to a schedule of 400 repeats even though he felt that doing a tempo run or longer intervals would have been more productive. He quickly learned to ask for the agenda ahead of time and do his own thing on such nights instead of staying with the group and wasting time.

But as Davis will tell you, despite the downsides, running with a group created a sense of accountability and community he came to value. "If I had to pick the best thing about running with the training group, it's that they pushed me," Davis says. "Yeah there were some petty annoyances with scheduling, but I really appreciated the mass energy when I was feeling fatigued and definitely would have backed off had I been alone."

SOLO OR SOCIAL?

Finding a partner or team to run with is easy enough. Local running clubs and athletic gear shops often hold regular meet-ups and training groups. There are a ton of apps and websites that do a decent job of matching up training partners, either in person or virtually. The trick is to get the right partner.

What makes a training partner compatible? The research seems to indicate that once you move out of the lab and leave virtual partners behind, the ideal running companion is someone reliable, slightly faster than you, and on a similar schedule to yours. Your partner should motivate you to push harder, but not too much harder. It definitely helps when you share the same work ethic as well. It's no fun waiting on the corner at 5 a.m. for a mate who shows up only about half the time. Or when she does show up, she's a mess without her coffee when you're raring to go; or she jumps out ahead like a jack rabbit while you prefer to ease into the miles.

That's a tall order, so it goes without saying you need to choose your running companions wisely. It's a good idea to chat about expectations—and I recommend thinking through a good breakup strategy just in case. The right partner can transform your running experience, taking your training and racing to new levels. The wrong partner can sabotage your performance.

Helen Cherono has a unique challenge when it comes to partnering up. As a blind runner, she depends on a running partner to "show" her the way. She and her guide runner are attached by a tether or string that's long enough to allow for arm movement but short enough so that when either one pulls on it, the other can change directions quickly; both the guide and the runner wrap one end of the tether around a hand and there's a knot at the center for an additional contact point. As the guide and runner move forward they can adjust the length and positioning of the tether to accommodate the situation. For example, in wide open spaces they can use the full length, so they can both swing their arms freely. When there's more congestion, the guide shortens the tether so he and his charge can maneuver through the crowd.

Obviously this is an extreme example of how the wrong partner can affect your run, but Cherono tells me it can be a setback if her match isn't up to the task. "The hardest part is when my guide runner is exhausted before the end of the race, or when the guide runner loses the string," she says.

Even if you usually run with a group, I still recommend an occasional run on your own. It offers some unique advantages. Whereas group running is ideal for dissociating some of the more laborious aspects, punching your own clock allows you to tune in and fine-tune; you can place all your focus on form, breathing, and the other technical aspects of performance. Five-time Olympian Francie Larrieu Smith backs me up on this. She says during her competitive years she would run alone when she wanted an introspective moment to pay attention to her form and pace; when she was looking to take her mind off certain aspects of her workout she would run with a group.

Solo running is probably the best opportunity for practicing associative thought strategies. You may recall how some runners are able to slip into a higher state known as flow. That's a strongly associative experience. Breaking away from the noise—and the pressure—of running with others seems to up your chances of achieving flow, especially if you're an experienced runner. (From what my runners tell me, that other higher state, the runner's high, can occur during either group or solitary runs.)

Running alone is certainly better than running with a mismatched partner or group. If you always run with a crowd that's way below your level, you will learn to loaf and probably lose some ground. Likewise, running with someone well beyond your abilities is a recipe for injury and exhaustion. Plus, I like the idea of preserving your own identity as a runner.

WRAP-UP

Running can be social or convenient, but rarely is it both at the same time. That's why I'm a big believer in the mix-and-match approach. Find a reliable partner or team and run with them, but only sometimes. Run on your own to preserve the solitude of running; check in with a group when you need a boost. Taking advantage of both the community and solitary dynamics of running allows you to understand and challenge your potential as runner without losing your identity in a pack.

CHAPTER 11

Your Racing Heart

Should you compete?

IF YOU'RE THINKING about starting a running program, why not jump in feet first? That's what Brit runner Alice Hampshire did. Her first run was also her first race.

"I thought, let's just see what happens," she says of the local 5K she popped into one Sunday morning about 6 years ago.

What made her do it?

"It was a whim. I'd been thinking about getting fit for a while, and the night before the race I thought, why not? So I went down the next morning and signed up about 30 minutes before the start. Half an hour later I find myself at the starting line with 300 other hardy souls. The gun goes off and then off I go."

What an experience! She managed to finish the course (flat and fast with a few rolling hills—sound familiar?) and averaged 11-minute miles, which on balance isn't terrible if you factor in a few quick recovery walks and one or two extra-long water stops.

Yes, she might have run a faster time if she had, you know, bothered to train. Yes, it would have absolutely been a better idea to set some goals, work toward them, and then try a race. And yes, her shins and quads were sore for a week. But those inaugural 3.1 miles lit a fire in her belly.

"I loved the feeling I get from running races right from the start. I was still soaking my bones in the tub when I signed up for my next one and started plotting my training," she says.

Hampshire is now a steady 20-mile-a-week runner and a regular pair of legs at 5Ks and 10Ks in her district. From where she started, she has improved immensely; training will do that. She's posting 8:45 miles with solid finishes in her age group (which she would prefer not to reveal).

Teddy Barrett, who hails from New Jersey, feels exactly the opposite way about racing. "What a pain," he says. "You gotta get up early, crowd in with everyone else at the start, and fight for a piece of the road until everyone spreads out. Get real."

As Barrett will tell you, racing is just not his thing. He's perfectly happy running alone or when the opportunity presents itself, going for a spin around the neighborhood with a few buddies. He's the guy who passes you on your morning run like you're jogging backward, so he's no slacker either. He's also no recluse. He simply doesn't love the idea of toeing the line with "300 other hardy souls"—or 3, or 3,000, or 30,000—to see where he stacks up in his age group or against the clock.

Here you have two dedicated runners, each with a different point of view about racing. As someone who works with runners, I don't find this surprising at all. Racing is one of those polarizing topics in the sport, like whether or not you should eat goo, or what is the best brand of running shoe. Your opinion is dictated by your goals, your personality, and your psychology. It's worth exploring whether racing fits into your overall Runner's Brain strategy or is perhaps something you should avoid like a bad heel blister.

SOCIALLY, RACING IS A MIXED BAG

While Hampshire is a social butterfly who flutters along with the crowd, Barrett prefers to flutter along like the lone leaf on a wintering tree. One thrives on competition while the other abhors it. Why?

As we learned in the previous chapter, training socially can teach you to work outside your comfort zone and push the envelope on performance. On the other hand, training solo teaches you to pay attention to your body. Both of these are important experiences to have as a runner, even if you prefer one situation to the other. But racing is a little different than training.

Racing is a unique cocktail of togetherness and aloneness. You're surrounded by runners on all sides plus the crowd along the sidelines, with communal mingling before and after the competition. There are also some unique social quirks associated with racing, too. How about those peculiar subculture bonding moments with strangers who happen to be standing near you at the starting line or running at the same pace? Or the war stories runners love to tell about that bathroom break at mile 10 or the time they surged past an arch nemesis with a hundred yards to go?

Yet with all this togetherness, racing is still, at its true heart, a solitary endeavor. Once the gun goes off, it's every runner for himself. You may be surrounded by thousands of people for miles and miles, but in the end you alone run those miles and you alone are responsible for your effort. It's up to you to push as hard as you can and post the fastest time you're capable of.

Perhaps you never thought about racing in this context before, though social factors have probably had more of a subconscious influence on your desire to race than you realized. Since we're talking about it now, though, ask yourself: Does racing's mix of social engagement and lack thereof please or annoy you? Does it inspire you or repel you?

REVVED UP TO RACE

Estelle Berkshire, coincidentally another British runner, tells me about one of the greatest runs of her life, a 5K where she shaved three and a half minutes off of her personal best.

"It felt amazing to have the motivation and drive to do it and although I still find it tough whilst doing it, the feeling after is pure elation. It may not be an impressive run to most, especially those running long distances, but everyone starts somewhere. And this is just the beginning for me!" she says.

Yes! Motivation. That's often the secret sauce for those who love to race. And as we can see on brain scans, motivation is more than just a psychological feeling. It's a feedback loop coordinated by many parts of the brain that goes something like this:

To start, a bundle of specialty areas housed in the front half of your brain light up, triggering an association of your original idea to "do something" with the concept of "achievement"—in this case, a race. Next, sparks fly as a shot of the neurotransmitter dopamine speeds down the neural pathways into the midbrain and the more primitive limbic system, tickling your gray matter with the urge to take action. Finally, the neurotransmitters zoom to the highly evolved prefrontal cortex. This is known as the executive center of the brain. It is responsible for decision making, reasoning, and action. When the prefrontal cortex lights up, it's the confirmation that yes, you should act on your do-something urges without further delay. The end result: You're fired up.

The brain can initiate these neural fireworks more quickly than it took to describe them here, or over any stretch of time, really. They can loop through the brain over and over again, or even bounce back and forth between the different phases. Or they can burn down to ash, in which case any sense of motivation disappears. The trick is to find an interesting enough target for your motivation—something with enough of a hook to keep the motivational feedback loop spinning.

Runners tell me there are all sorts of reasons that keep them racing. As I told you in Chapter 4, goals are often the ultimate motivator for running in general. That is especially true for racing, I think. For some runners, having a race in their sights is the only reason they need to step on the gas. Gunning for a PR is what gets some people out of bed early in the morning and provides a kick in the shorts when they're feeling lazy. A race penciled onto the calendar is the incentive they need to log those few extra miles when they otherwise wouldn't. Like a running partner, a race date equals accountability. One runner told me his goal for entering races was to build his T-shirt collection—I'm not kidding! If that's what gets you out of bed in the morning, why not?

For others, the inspiration for racing comes from the same place that compels them to track their steps, minutes, and miles with health-tracking devices and apps; they're the wonks who dig having data to analyze. For someone like that, racing provides the bricks needed to build the structure of their training. More importantly, it becomes a measure of progress. Because the clock is finite and objective, it draws a straight line between where you started and where you are. A minute shaved from your 10K PR is irrefutable proof you're getting faster. The brain scoops up this information and feeds it right into the motivational loop to keep you going.

Not everything is about personal bests and exploring your limits, though. You will find plenty of runners on the racing circuit these days who aren't speed demons and who couldn't care less about clocking in a personal best. They race for a reason that's near and dear to my heart: charity.

Many racers with a heartbeat for giving back wouldn't have gotten out the door for a single training run unless they felt a need to serve the greater good in some way. I think it is an amazing way to combine doing something for yourself with doing something for others.

Stephen Liegghio tells me his greatest run was the Brooksie Way Half Marathon in Oakland County, Michigan, in 2012—which he ran 45 days after donating one of his kidneys to a complete stranger. It wasn't his fastest half, but he wore a shirt letting everyone know he had just donated.

"It's the most support I have had of any race," he says. "People were stopping me during the race to take their picture with me. I felt like I really inspired people that day."

Likewise, Michelle Jones says her first Racing for the Cure experience was truly memorable. "I had walked it the previous year with my daughter in a stroller and my two-year-old son holding my hand. I was sick, bald, and still undergoing chemo, but I told my family the next year I'd run. And I did," she says proudly.

REASONS NOT TO RACE

Racing is every feeling you have as yourself as a runner squeezed into a ball. It can stir up a brew of thoughts and feelings about everything from the months of training to crossing the finish line. You learn something about yourself. Like, you're faster than you think. Or stronger, tougher, and more resilient. Just the decision to race can bring those things out in you because they change your self-perception. I've had runners tell me the moment they began to think of themselves as someone who enters a race is the moment they began to build an identity as an athlete.

Yes, racing can be a great passion for a good many runners. But your mom was right when she said jumping off a bridge isn't for everybody. Just because all your friends are doing it doesn't mean you have to toe the line, too.

Plenty of runners I have worked with steer clear of races their entire career specifically because of the social aspects. They get wigged out standing at a crowded starting line, or weirded out when cheered on by strangers. Sometimes they prefer to keep their pace private. Or maybe they just view running as a personal meditation. Whatever. The racing scene isn't their bag.

You don't plan your vacations around the marathon majors? That doesn't mean you aren't committed to your running program. You may be striving to go farther and faster, but at least for now you don't wish to make any bigger commitment to your training program than you already have. Like our friend from New Jersey, you may not want your time taken up with prepping for the race, fighting through the start, or waiting around afterward. Even the act of pinning a bib on to your T-shirt may seem like just too much of a hassle.

That's okay. Your head may not be in the right space to race. It may never be. Running a race requires concentration, not just on race day itself, but also for the discipline of training and living a certain lifestyle. If it's not for you, then it's not for you.

I have seen this sort of attitude in people who run a bunch of races until the mental fatigue finally catches up with them. Maybe they popped a bad one last time out and need a timeout. These are the folks who may run some races again at some point but need a break for the foreseeable future.

I have also known some newbies who might make an appearance at a starting line at some point in their career but just aren't ready yet. Either they don't have enough miles under their belts, they lack confidence, or they just have other priorities at the moment. I say, no point in rushing things. You will either get there or you won't. No need to feel guilty or less than. The world keeps spinning either way.

WRAP-UP

Entering some races or even getting serious about competing on a regular basis is a rite of passage every runner at least thinks about. But racing isn't everyone's jam. If you love the thrill of throwing your arms up for the picture they take at the finish line, that's cool. If you prefer to avoid the crowds for social reasons or for performance anxiety—also cool. Either way, the miles still count.

CHAPTER 12

Managing Competition

Overcoming pre-race jitters and post-race blues

ASSUMING YOU HAVE decided racing is for you, I've got a bit more psychology to share with you.

What I find is that most people within a race do pretty well. Once the gun goes off, it takes only a few seconds to focus on the task of running the race. Occasionally runners will hit the wall or some other sort of snag, but for the most part they either tune in or tune out to get through it. Psychologically speaking, most of the trouble occurs before and after the race.

In this chapter, I explain what those "pre-race jitters" and "post-race blues" are all about, then give you some strategies on how to deal with them. These stategies will help you transition seamlessly into and out of competitions.

PRE-RACE JITTERS

Let's say you have selected an event, signed up, circled it on the calendar, and trained your heart out. You get to the start and you're so nervous you feel like you might pass out. Intellectually you know it's only a local 5-miler, but emotionally it feels like an Olympic final.

Back in 1972, Kathrine Switzer became the first woman ever to officially enter the Boston Marathon when it was still a men's-only race. (She ran it unofficially in 1967.) She went on to run more than 40 marathons, winning New York once and finishing second in Boston several years after the inaugural run. She's about the smoothest operator I have ever met, but she admits she used to get so nervous before a race she almost couldn't function.

"I figured this out when getting nervous was just crippling me—it was the on-time thing," she says. "The thing that makes me the most nervous is that I had to figure out getting to the race on time and getting settled and warmed up and using the toilets in plenty of time."

Obviously, some runners feel the pre-race jitters more profoundly than others. But I think all runners feel a little bit of flip-flopping in their stomach before a race. It comes with the territory. A little case of anxiety before a competition may actually be a good thing in terms of performance. To understand why, let's reach back to 10th grade biology for an explanation.

If you were paying attention in class (and have a good memory), you may recall that your sympathetic nervous system (SNS) controls your heart rate, blood pressure, and blood vessels. That heightened, nervous feeling you get just before the gun goes off or if you unexpectedly meet a bear in the woods is caused by the SNS pumping adrenaline and other hormones into the bloodstream. When the agony of waiting through all those pre-race announcements is finally over and the gun goes off, this flood of "fight or flight" hormones starts your heart racing and causes your blood pressure to climb, leaving you feeling instantly more alert and primed for action. You take flight, so to speak, thanks to that extra shot of energy to propel you forward.

If you're super keyed up, however, it can be an issue. In an overly agitated state you're likely to rocket out too fast, leaving nothing for the later part of the race. On the other hand, a lack of adrenaline rush is also trouble; you won't have the juice to push your performance. (I don't know too many runners with this problem, but I will occasionally see someone so relaxed they barely react when the horn goes off.)

Pre-race planning helps strike the right balance between the SNS response and keeping your cool. Just as you have done your physical prep for the race, some mental prep is in order as well. I've got some strategies to help you do that.

Calm the Jitters Strategy #1: **Visualize**

Remember how in Chapter 5 I told you the story of Mark Plaatjes's amazing World Championships marathon run in 1993? Plaatjes was the master of visualization. By the time he actually ran the race, he had envisioned the event so often he knew every twist and turn, every crack in the pavement, every leaf on the trees that lined the course. Plaatjes was able to channel his nerves through mental imagery. His familiarity with the race, even if it was only in his mind, went a long way toward helping him relax. I'm sure he got the pre-race pumps. He was able to control it by knowing his plan for every step of the race.

Some runners benefit from tapping into actual past experiences. Think back to your best race or a strong training run. Now take those feelings of past success and own them for the present. Runner Bert Rodriguez says he channels past events all the time to help him through races.

"Whenever I'm feeling tired I remind myself I once ran four loops around Central Park. That helps me put things in perspective. If I can do that, I know I can handle the miles I've got in front of me right now," he says, adding that as soon as you learn to trust your training you start believing "you got this."

The moral of the story is that you should attend to the things you can control and minimize the things you can't. There's nothing you can do about the weather, the crowded field, or a lack of water stops. You can only control your reactions. If you're properly prepped and you have done both your mental and physical homework, you have less to worry about. Switzer says that as nervous as she gets before a race, the one thing she never worries about is her training. There's no point in that, she says, because by race time it's too late. Her philosophy is that by the time you get to the starting line, you have done everything you can and you just have to go for it.

Calm the Jitters Strategy #2: **Positive Self-Talk**

Another thing that helps cut through the jumpiness is positive self-talk. When you come down with a really bad case of the yips, it means fear has set in. Fear throws off your pacing, it makes you doubt your strategy, and it messes with your sense of energy management. That's when you start making mistakes. Doubt cracks open the door to failure, and second-guess syndrome never did anyone any good.

Your goal should be to manage your anxiety by thinking confident thoughts. Try coming up with some canned mantras you can easily remember and repeat to yourself. Runner Jonathan Labell says he likes to invent positive statements for himself like: "I'm strong, I'm fast, my speed will last."

"It's catchy, it rhymes, and you can conform it to the rhythm of your stride," he tells me.

Mantras that take potential negatives and turn them into positives also work. Instead of telling yourself "My quads are killing me," you could say "It hurts so good" or "It's only temporary."

I recommend reviewing all the information I gave you in Chapter 4 about goal setting. Consider setting a series of what I like to call your best, great, and good goals. Your best goal might be to hit a per-mile pace for the entire race. Your great goal might be to run a negative split between your last mile and second to last mile. Your good goal might be to simply run the race as fast as you can on that day. If you flex your expectations, you can enter a race knowing you will feel good about what you accomplish no matter what.

Labell, who takes this positive psychology stuff seriously, tells me he writes his pep-talk phrases on the inside of his arm or along the side of his shoe. But if you don't feel like defacing your person or property, you might write a note on the back of your bib or on a slip of paper in your pocket.

Calm the Jitters Strategy #3: **Don't Overthink**

You can make yourself crazy with splits and pace and course features. Just try to come prepared and give the best effort you can give on that day. By the same token, you want to avoid letting worry blow the physical side of things. Be careful about overtraining the week or so before the race. Avoid overeating due to over-anxiety. And don't leave your race at the start by going nuts on your warmup.

Calm the Jitters Strategy #4: **Be Superstitious**

One last thing that may help keep you calm on race day: Stick to your rituals. The lucky underwear we talked about in Chapter 8? Wear it if you got it. Magical thinking, so as long as it doesn't get out of control, can be the best soothing strategy of all.

POST-RACE LETDOWN

Once the journey is over and you have told your tales of victory and defeat over a few post-race spritzers, you may be left feeling a little lost at sea. After training, thinking, and planning for so long, what do you have left to look forward to?

I know a runner—triathlete, really—who way back when decided to train for a double Iron Man. In case you aren't familiar with the math, that's 4.7 miles of swimming, 220 miles of cycling, and 52.4 miles of running.

Those distances are no joke. You have to train really hard to complete them, and train she did. She worked her tail off for over a year, running mile upon mile, swimming laps in the pool, and grinding out the miles on her cycle. A typical weekend involved running a marathon on Saturday and doing a single-digit Iron Man race on a Sunday. It took all-consuming dedication to keep going, not to mention every drop of free time she had.

Finally, race day came. She rose to the challenge. She finished up the event in around 28 hours, and placed seventh among the women and 30th overall. What an amazing feat!

Then she didn't run again for 7 years.

You could say that this runner (who prefers not to be named) is a woman of extremes. That's probably true. But it's also true that after an event she had dedicated her life to was over, she had a serious case of the post-race blues.

Her analysis of the situation is that she no longer had a reason to train. She wasn't considering another double, and anything less just didn't seem worth the effort. She did explore the idea of running a triple but ultimately felt like that was a bit much, even for her. It took 7 years before she started thinking maybe shorter distances could be a different type of challenge for her—speed versus distance. When she started back in training, she traded her LSDs (long, slow distances) on the road for sprint ladders on the track.

Hitting your goal by crossing the finish line is an amazing feeling that is often followed by some serious letdown. Completing a double Iron Man—or any other distance you have seriously trained for—can leave you asking the question: What purpose does my training now serve?

These feelings may catch you by surprise, but they certainly aren't unique to runners. Athletes from every sport at every level report feeling the same way after an important event. Politicians, students, surgeons, lawyers, and anyone achievement oriented are susceptible to a sense of aimlessness after completing a major goal, regardless of whether the outcome was win, lose, or draw.

While finishing up something important like a race is a good time to reflect on your accomplishments, post-race blues are perfectly normal. Your training filled your life; now you have to fill the void, logistically and emotionally. Switzer said when she was running competitively and only did two marathons a year, her whole world was focused on racing and she would get very blue about 5 days afterward.

"This is because I put everything into it. I was physically drained and had changed my whole body chemistry as I was depleted. Also, by then, the endorphin high of the race was all gone and it was like a crash!" she says.

Just as there are things you can do to calm your nerves before a race, there are things you can do to lift yourself out of the post-race blues.

Pick-Up Strategy #1: **Enjoy Your Break**

Switzer says that once the race is over, you need to do something else for a while that gives you a sense of fun and accomplishment. That's good advice.

Before a race, you have pressed the go button for so long that it's hard to power down once it's over. But that's exactly what you should do for a while. I know plenty of runners who experience a vague sense of guilt when sleeping late after a race, but R&R is exactly what you need. Even if you have something else big planned in the future, it won't set you back if you take a little time to pay attention to other things you neglected in life while you were consumed by your goals.

So, yes, turn off the alarm clock for a few mornings. Hang out with friends. Catch up on your reading. In general, recharge before you hit the road again. There's a good chance physical soreness will keep you from doing any meaningful workouts for at least a few days anyway.

All that said, you don't want several days of relaxation turning into 7 years, as with she-who-asked-not-to-be-named. Just as you planned and plotted for your big race, you should plan and plot your post-race training. That's not to say you want time trials and tempo runs baked into your training right after finishing a race. You just want a plan to continue.

Don't let too much time go by without lacing up, even if you just get out there to run for the sake of running. Having a post-race plan can help with the post-race blues that I talk to so many runners about. Keep in mind that before a race, life is highly structured. After a race, the structure disappears. It's important to have scheduled activities, even if it's not running. Renew neglected relationships, do something with family or friends, write down your experiences, and start investigating your next race.

Also, if you have stopped running and were benefitting from the feel-good chemicals feeding your brain, those too may be changing, and you may feel more blue as a result. If the post-race blues stay around (sometimes they do if they are paired with an injury, significant performance decline, or debt because of training, travel, and missed time at work), seek out support from a professional who knows how important running can be psychologically and socially.

Pick-Up Strategy #2: **Reflect**

Make sure to take a moment or two to reflect on your performance. Consider what you did well and what you could have done better. Even if you aren't entirely pleased with your performance, don't throw yourself a pity party. Think about what worked and what you can do better next time. However it went down, it's worth analyzing why things went the way they did. You're stocking up information for the next time you're standing at the start of a race with a case of the jitters. Now you've got some past life to relive, some manufactured fodder for positive thoughts or better planning.

Avoid the trap of letting others bring you down, especially if you're already feeling a tad disappointed. If someone asks about your time, tell them you finished and it was awesome. If you didn't finish, tell them you're happy with your effort. Don't indulge in comparisons that leave you feeling like you somehow came up short. You had the courage to try. Never forget how amazing that is.

Pick-Up Strategy #3:
Set New Goals

After you take a pause, consider hitting the gas pedal again. Unless you're burned out and have decided not to race for a while—or ever again—pick a new goal. Dedicate yourself to something new. If you're bored or burned out, maybe set your sights on a different distance, a new kind of course, or even a different activity. Nothing stokes your love of the road like a passion project. And a fresh perspective is never a bad idea.

Pick-Up Strategy #4:
Gratitude

Remember, too, to give thanks. Behind every great goal there's a supporting team: a wife who got up to make breakfast for you before those crack-of-dawn training runs; a coworker who took up a little slack so you could make an early workout; a child who drew you motivational posters. Nothing helps you snap out of a post-race funk like the realization you have a lot of people rooting for you.

WRAP-UP

A certain amount of pre-race jitters is normal and even desirable. But if you let them ratchet up too much, it might influence your performance in a negative way. Why deal with the misery when there are ways to dial down the emotions? Use the strategies I suggest in this chapter to keep calm and carry on before, during, and after a race.

CHAPTER 13

Mental Blueprints for Some of the World's Toughest Races

Straight from the horse's mouth

SOMETIMES A RACE is just a race. But sometimes a course is so steeped in folklore it takes on almost mythical proportions. When race details are printed internationally for runners to hang on every word, those who have already run are peppered with a thousand questions by those who want to take the challenge. Even if you don't live within a thousand miles of a particular race, you can recite minute details about the route, the history of it, war stories from the trenches passed down over the years.

The four races I detail in this chapter have achieved just such iconic status. They are among the greatest beaten paths many runners dream of tackling someday.

If you decide to take on one of these events, good luck. Train well. Put your time in on the track and on the road. Hit the weight room. Stretch out your hamstrings. But don't forget the importance of training your mind to go the distance. Myths and folklore evoke fear, like ghost stories around a campfire. By now you know mental prep is every bit as important as physical prep.

To help your brain get a leg up on these competitions, I have spoken to the people who know them inside and out. In this chapter, they share their insiders' points of view to give you the scoop on the best strategies for running your best race from a mental perspective. What's really awesome about their advice is that it is specific to each course, yet transcends the particular race so it's applicable to practically any race or training run you encounter.

So read on. Soak up the knowledge that will give you the psychological edge to run your best race, whether it's the dream event you hope to run, just a local road race with a similar profile, or anything in between. It's nice to know that the strategies I have given you in this book truly do have real-world applications. Just ask the pros I have spoken with—if you can catch them.

THE ICONIC 5K: CARLSBAD 5000

Billed as the "World's Fastest 5K," the Carlsbad 5000 is open to competitors of all ages and fitness abilities. There are separate races for men and women, which are then divided by age groups, and separate heats for walkers and wheelchair competitors.

The tall, raven-haired Grace Padilla has been a regular at the starting line year after year. She is so gorgeous she should be on the cover of magazines—in fact, she has been on the cover of magazines. But make no mistake: Padilla is always a consistent Carlsbad contender. Now a masters runner, she won the race once in her twenties and has finished near the top of the standings all eight times she has run it.

Padilla acknowledges the course is fast and flat but says it pays to know the route so it doesn't throw you any surprises. For example, watch out for some rolling hills in the second mile. They will sneak up on you if you're not careful. Also, as with any 5K, Padilla says that to run a good time you have to adjust your thinking. She changes the language in her self-talk to accomplish this.

"To run longer races like a marathon, it's all about enduring time," she says. "For a 5K, it's all about enduring speed."

The unique challenge of the 5K distance is that it's too short to be considered long distance and too long to be considered a sprint. You have to build up a tolerance for high-intensity work, something that is uncomfortable no matter how well trained you are. Getting psychologically prepared for such a task is not something you want to leave for race day. Accepting the uniqueness of a course like the Carlsbad gives you plenty of psychological wiggle room when you need to adjust your thinking to match the demands of the course.

When she's training for Carlsbad, Padilla said she often feels insecure before a training run. Most likely, the self-doubt comes from what she's predicting about the run, probably something negative. She has trouble wrapping her mind around how much work and pain she is about to put herself through. Once she thinks about what she wants to accomplish, she notices how it changes her thinking.

"But then I warm up and I tell myself to take it one rep at a time," she says. "I amaze myself with what I am capable of doing. I spend enough time building up confidence in training and it translates to self-assured thinking on race day. You will believe you are more courageous if you have faced down a fast mile or two and lived to tell the tale." Now that's what you want to tell yourself about a tricky course.

Learning the art of pacing is also an essential 5K skill. Padilla says that runners who lose emotional control and get overly excited go out way too fast, and by mile two they're spent. But if you spend time in your workouts learning the emotional feel of the race pace, you shouldn't even need to see your split times to know how fast you're going.

As for the race itself, Padilla offers a couple insider tips.

First, it all comes down to focus. If you can find that sweet spot between discomfort and rhythm, you have a better chance of getting into the kind of flow state we discussed in Chapter 7. Then any pain you feel will disappear and the race will seem like it's over in a flash.

Second, don't let other runners distract your focus. Sticking to your race plan is really important at Carlsbad or any other 5K because the race is comparatively short. You don't have a lot of leeway to correct your mistakes. If you let the emotional surge at the start carry you faster than you should be going, you will implode before the halfway point; emotionally and physically you will be too fatigued, setting yourself up to make more cognitive errors about the remainder of the course and how you should handle it.

Padilla likes to pick another runner and sync herself with the steady motion of their legs, or concentrate on following the center of the road, which is generally the shortest distance to the finish line compared to the inside and outside edges. Psychologically, kicking from behind seems to be easier than trying to hold off a pass for most runners.

THE ICONIC HILL: MOUNT WASHINGTON 7.6-MILER

Every year, on the third Saturday in June, over 1,100 runners gather at the base of Mount Washington in New Hampshire for a 7.6-mile "run to the clouds." The course features only one hill. The phrase "one and done" takes on a whole new meaning.

From base to summit, the course climbs 4,500 feet with an average 12 percent grade along the Mount Washington Auto Road. Think about how steep a treadmill is when you set it to 12 percent grade. Now imagine chugging up that incline as fast as you can for 2 hours straight with no break. Add in fog and a 40 mph wind, or 70 degrees and a beating sun, or a 30-degree differential from top to bottom, and there you have Mount Washington.

"The race is whammo in your face from the first step," explains John Stifler, the race's official spokesman. "It just goes straight up. In that sense it's as hard as any run there is anywhere."

Fighting and winning the uphill battle that is Mount Washington depends on physical training, of course. But Stifler says in many ways, it is largely a psychological challenge: If you don't approach it in the right way, you're more likely to end up with a crushed spirit rather than a pulled hamstring.

Stifler is certainly the guy to give you the inside line on the brains of this race. He has been involved in the running, planning, and management of the event in one way or another since 1989. If anyone knows how to approach this course, he's your runner. Some of what he has to say is counterintuitive.

"Many people do well the first time they run the Mount because they have had the fear of God put into them. They have heard the horror stories about the continuous climb and as a result they tend to start out more conservatively. So in a sense, the fear of the unknown can serve a runner very well," he says.

The ones who avoid an overenthusiastic start the first time around sometimes get into trouble when they try the race again. They train for it even harder, thinking they will push their limits and go for a better time. But often they do worse because they use up too much in the first half of the race and run out of gas for the second, Stifler says.

According to Stifler, no runner he has ever talked to wishes he had gone out faster. That includes Olympic gold medal marathoner Joan Benoit Samuelson and three-time race winner Eric Blake. Samuelson told me it's one of the hardest races she has ever run—and she's an Olympic gold medalist! Even elite, seasoned runners like Samuelson regret being lulled into a false sense of security by the foreshortened first mile and the slight downhill right at the beginning of the course.

The best way to run the race? Stifler advises sticking with a rational strategy that relies more on your brain than your gut. "Figure out a detailed plan before you run the race and don't deviate," he says. Luckily for us, he has come up with a formula he says should help a lot.

First, take your usual per mile pace and toss it out the window. Now, take your average half-marathon time and divide it by 7.6. That should be about your average overall race pace.

Since many runners don't realize that mile markers are the permanent road signs rather than precisely measured distances, some don't know that the first mile is actually a little short because it's measured from the end of the parking lot rather than the starting line. So Stifler says to anticipate a first mile that's around 90 percent of your average overall race pace. Also, plan to average a minute per mile faster in the first half of the race compared to the second half.

If you have never done the race, Stifler says to prepare yourself mentally for the perpetual climb. "It's not like other hilly courses where you're used to thinking, 'I'll get to the rise and take a breather,'" he says. "There are no breathers here, no top of the next hill. You just keep going up."

Stifler also warns that you will probably be in a lot of pain and think about bailing at some point during the second half of the race. That's understandable, but you have to keep your brain and body in sync. Consider using some visualization techniques to do some dry runs of the course. As you'll recall from Chapter 5, imagining a run is like giving yourself a dress rehearsal for the real thing.

"Just remember those times you felt that way in other races only to realize that running, like childbirth, is 'the kind of pain you forget,'" Stifler says.

To get through the pain, Stifler advises continually congratulating yourself for the fact that you are still moving forward in this moment, no matter how slowly you go. "Tell yourself that if you keep doing what you're doing, you will get to the end. If you can just focus on that thought, you can make it," Stifler says. "When it's over you will say, 'I'm glad I did it.'"

In general, Stifler says distraction and dissociation strategies (like the ones I outlined in Chapter 6) should serve you well, too. Think thoughts that distract you from the pain rather than allowing the pain to overrun your thoughts.

As for walking some of the race, Stifler says there are two philosophies. One school of thought says walking is okay because it helps you gather your strength, and you will probably move about as fast as you run anyway. The other school of thought says each time you shift to a walk, mentally it gets tougher to hit the gas again. If you do decide you're okay with some walking, Stifler recommends planning your walk breaks, both when you're going to take them and for how long. If they are intentional, you won't feel defeated because you had to slow down. This way you are sticking with a plan, which can help you feel confident that you are accomplishing what you set out to do.

And once you do complete the race, Stifler says the experience can help you during tough points in other races. "After you have done Mount Washington, you will never again be afraid of any hill in any other race ever again," he says. "It's like having a black belt in running."

ICONIC HALF-MARATHON AND MOUNTAIN RACE: PIKES PEAK

Ron Illgen, president and race director of the Pikes Peak Ascent (13.3 miles) and the Pikes Peak Marathon, likes to say that both of his races start out so sedately you feel like you're running down a street in Anytown, USA.

But then you reach the end of the first paved mile and take a left turn. "At this point, any illusions you had that this will be a walk in the park should vanish," he says in somewhat of an understatement.

As soon as you round that corner, your feet immediately point upward—and stay that way for the next 10 miles. When you finally reach the timberline at 11,000 vertical feet, you are up so high even the trees have the good sense not to grow there. And from there you still have 3 miles and 2,000 vertical feet to go until the summit.

Adding insult to injury, it might start to snow, and there might be a 30-degree differential from base to summit. Along the way, protruding rocks and roots are waiting to send you crashing to the ground, lacerating flesh and only temporarily masking the pain of blood-filled blisters. As Illgen describes it, when you take on Pikes Peak, your legs, lungs, heart, and mind will be worn to a ragged nothingness. Sounds like fun, no?

If you are running the Ascent, your work is done once you tag the top of the trail. You can claim your medal, take a bus back down to the base and start to nurse your beat-up body back to health. For those braving the marathon distance, the summit is only the halfway point. From there you will turn around and descend the same path from which you came. That's more than 13 miles straight downhill. (More about that in just a sec.)

Still want to take on this race or one like it? Illgen says it's possible to survive it and maybe even thrive, so long as you do your homework and have mad respect for the undertaking. A lot of his advice is similar to what should work for you at Mount Washington: Train well mentally and physically, have a deep knowledge of the course, and adjust your cognitive expectations.

Illgen says the race office puts out a step-by-step description of the course. Study it. Knowledge is not only power, in this case it's also safety—and probably the difference between finishing and not finishing. Definitely don't attempt the course blindly. In fact, if at all possible, Illgen recommends competitors get out to the Peak a week or so early to hike parts of the trail and get familiar with the terrain. Stepping on to various parts of the course can inform your visualization practice and also give you a chance to think up positive mantras for obviously tough spots. This can also help lessen any pre-race jitters you might be feeling.

The most important piece of mental strategy runners can carry with them up the mountain, Illgen adds, is to focus on landmarks rather than per-mile splits or the race as a whole. "The mountain is so beautiful, but whatever you do, don't look up ever," Illgen warns. "If you allow yourself to face the task of climbing the mountain as one, large undertaking, you'll feel discouraged."

It's advisable to break up your main goal, which is finishing the race, into a series of smaller goals using some of the goal techniques I outlined for you in Chapter 4. With such a steep climb and wildly variable terrain, consider each mile as a goal unto itself. Also, expect a 20- to 30-minute differential between your fastest and slowest mile splits. Plus, it's more fun to have several accomplishments in one run than only one accomplishment at the end. Boston winner and Olympian Meb Keflezighi has a whole philosophy about this. In the book *Meb for Mortals,* he writes: "You should have several goals going into a race. Your list should start with your ultimate goal, and work downward from there to several potential outcomes that, while not your ultimate aim, are still worthy accomplishments."

At some point, you will almost certainly be forced into a zombie shuffle. The combination of distance, bumpy trail, and elevation means that more than 95 percent of competitors will walk some of the time, if not for a majority of the race. Here Illgen's advice on walking differs from what Stifler recommends for Mount Washington. He says there's no shame in necessity.

"Run when you can, walk when you can't" is his straightforward advice.

A word for those running the marathon distance: Once you make it to the top, the hard part is yet to come. The biggest mistake people who run the marathon is underestimating the downhill, Illgen says.

Downhill running trashes your legs and does a number on your head. Because it puts your muscles, particularly your quadriceps, into a perpetually lengthened position, extended periods of downhill running can be excruciatingly painful. You may even begin to feel the soreness you typically don't feel until after the race while your run is still in progress. If you get too focused on the pain, you won't pick up your feet properly. When your feet don't move well, you raise your risk of stumbling over rocks and roots along the way. Illgen says that every year the race sees its share of broken bones and lacerations, especially on the way down.

Since there is very little you can do to prepare yourself physically for the downhill pain, you have to do the best you can to prepare yourself mentally and emotionally. Illgen says many of the competitors he has spoken with seem to do best with dissociative strategies. Because there is so much discomfort concentrated in the front of your thighs, it may be too much to ask your brain to associate into the pain and embrace it. Most people seem to tolerate it better if they can get out of their body and focus on something else.

THE ICONIC MARATHON: BOSTON

The Boston Marathon is like the World Series, Super Bowl, and Kentucky Derby of marathons. It's the most prestigious running race of any distance in America, perhaps even the world. For marathoners everywhere it's the Holy Grail, the one they hope to run at some point in their careers.

For more than 13 years I have had the honor of being the official medical team psychologist for Boston. With up to 36,000 competitors toeing the starting line each year, there is plenty to keep the medical team, including me, busy in the medical tent. When you hear about planning for the Boston Marathon, you're going to hear race director Dave McGillivray's name come up over and over.

McGillivray not only is Boston's race director, he also has run the course 41 years in a row. He was an official entrant for 15 years. Then, once he became the director, he began driving to the starting line several hours after the majority of runners finished the race to run it himself. Somewhere around midnight on race day, he is now the last runner to complete the course.

McGillivray's "Pahk the cah at Hahvad Yahd" accent lends an air of authenticity to any advice he offers about running Boston. He certainly has lots of advice to give.

"There are effectively three challenges and three types of pain to running Boston," he says. "Physical, mental, and emotional."

In the past, marathoners largely focused on the physical and mental aspects of the race, McGillivray figures. They would train hard, and they would race hard. They knew they would cross the finish line. It was just a matter of how fast they would run and how many people they were going to beat. Today there are more people participating in the marathon than ever before, McGillivray points out. The walls of intimidation are coming down, in large part because philanthropy organizations like Race for the Cure and Team in Training have entered the mix. Many runners now run for a higher purpose, to raise money for their charity of choice.

That's where the emotional part of the equation comes in.

"I really think that a lot of people now have a deep emotional connection to racing and training that helps get them through the difficulties," McGillivray says. "They might feel really bad along the course and consider bailing out, but then they remember the individual or the cause they are running on behalf of and something happens. The adrenaline kicks in and they're able to continue."

No doubt, emotion can be a powerful motivational driver. You may recall from Chapter 4 that the first tent pole in the goal-setting system I gave you was "feeling your goal in your gut." Working toward something you feel passionate about achieving helps you connect with your objectives on a deeper level. And what tugs at your heartstrings more than a cause you truly believe in?

McGillivray also thinks that when running Boston (or any marathon), physical strength is inversely proportional to psychological strength. In other words, you start out physically strong but a little unsure about your prospects. As the race progresses, your physical strength wanes but your confidence builds. Why? Because as the miles click by, even as the lactic acid spreads through your muscles, you move closer to the finish line. And the closer you move to the finish, you become surer you will actually make it.

Most people think that mile 21 or 22 will be the toughest point of the race. Not so, says McGillivray. Interestingly, he says for most runners he knows, it's around mile 10: "At this point you have run a long way but you still have more miles in front of you than you do behind you. That's a tough mental challenge for a lot of people."

Though many runners fear hitting the wall in those later miles, it's an unlikely occurrence if you have trained properly. McGillivray says that, paradoxically, it's the point where the race can actually get easier for a lot of runners because you have only a few miles left. Even if you have to stop and walk the rest of the way, you would still make it across the finish line. Mentally, that sets you up for success. So if you can make it to the part of the race where you are counting down the miles rather than counting up, you will probably feel physically spent but psychologically sturdy. (Flip to Chapter 14 and read up on how to push past a wall, just in case.)

Finally, before we leave Boston behind, a few thoughts about the infamous Heartbreak Hill, which McGillivray hesitates to say is overrated. He says you need to think about it as just another hill in another race. However, given the history and tradition of this particular uphill battle, this is what McGillivray says can happen:

"The thing about Heartbreak is that it's fourth in a series of four hills. The first is a grade that starts at mile 16 and runs over Route 128. It isn't steep but it seems to go on forever. Some people think that's Heartbreak. Then, the second hill hits once you turn the corner at the fire station at 17.5 miles. Someone not familiar with the route might think, what is this? I've already run Heartbreak.

"Then, another mile down the road, you hit yet another hill. You climb that and think this has got to be Heartbreak. You get that over with and all of a sudden—mile 20! And here comes Heartbreak.

I think that's what breaks your heart. You think you have already done Heartbreak two or three times, and here it is again. It's like Groundhog Day."

So clearly the best mental strategy for tackling Heartbreak is something you have heard before: Know the course! Also, relax. McGillivray says if you have done your homework, getting up Heartbreak will be easier than you think. In fact, a lot of people ask him after the race where Heartbreak Hill was. He thinks one of the reasons for this is crowd support.

"It's almost like camouflage, like it's buried. Of course you see the hill, but it doesn't look ominous like if no one was there and it was barren and all you were doing is staring curb to curb at this very steep hill. You look and just see people and it takes your attention away. You don't focus on the climb as much," he says.

WRAP-UP

So now you have strategies for some of the most iconic road races around. It's easy to see how well they translate to other running challenges, too. Are you inspired to give any of these greats a try? Now that you know more about how to approach them, they're waiting for you. See you at the starting line.

PART 4

CHALLENGES

AS A RUNNER, you're bound to experience great highs and depressing lows. When you set a new PR, conquer a new distance, or discover a new running trail—those are the highs, and they are awesome. But let's face it: Lows are no fun. Those days when a three-miler feels like an ultra? Or you're at the far end of an out and back and your hamstring starts to complain? Or you're running your heart out and a petite lady pushing a baby stroller goes zipping past you? On days like these, it feels like the weight of the world is in your running shoes.

In dealing with runners for more than a decade, I have learned a lot about the obstacles they frequently face. Some are borne of physical exertion, such as injuries or hitting the wall. Others come from life circumstances that inevitably rear up and get in the way. (Anyone who has had a child or checked the box on a race entry form for a new age group knows exactly what I'm talking about.) And others are strictly your own creation, spawned in your mind and taking on a life of their own.

In previous sections I have given you some tools for smashing through mental barriers. Now we're going to take those tools and apply them to specific challenges almost every runner I know has faced or will face at some point in the future.

CHAPTER 14

Hitting the Wall

Running's great equalizer

TEXAN STEVE CROSSLAND can vaguely recall his first marathon. He's got a fairly clear memory of everything that happened early on in the race, but then at mile 23 he blacked out. Everything is "fuzzy" from there on in, he says. His friends tell him he stopped cold at mile 25 or so, and then the EMTs came to assess him.

"I literally just stopped and started walking unsteady . . . I was answering questions from the EMT. 'What's your name?' 'What city are you in?' 'What mile do you think you're at?' All of which I knew, but I couldn't stand without assistance."

Crossland was so wobbly, weak, and disoriented that he was given a ride on the back of a Gator utility vehicle to the medical tent. After an hour or so of being pumped with a bag or two of saline, he felt fine. Unfortunately he never made it to the finish line the old-fashioned way.

In runner-speak, we call what happened to Crossland a classic case of hitting the wall. It's that dreaded fatigue that brings runners into the medical tent. It can be so extreme that your heart pounds raggedly in your chest, your muscles shake uncontrollably, and every step is a triumph of will. Legs start feeling like concrete posts, and you seriously doubt the race actually has a finish line. Cyclists refer to it as bonking; the Brits sometimes call it "hunger knock." By any name it's a pretty awful experience.

Most marathoners report that they usually hit the wall somewhere between miles 16 and 24, though some tell me when they have had a really bad day they wind up crashing at a much earlier mile marker. Anecdotally, the estimates are that about half of all non-elite runners hit the wall at some point in their racing careers. I have never seen a study to substantiate that stat, but in talking to runners both inside and outside the medical tent that sounds reasonable, especially as more and more casual runners enter long races with little knowledge of their bodies and the course.

As much as I would like to say it ain't so, even elite runners aren't immune to the wall-crashing experience. That includes Olympian Meb Keflezighi. He talks about a couple of times he had been in major marathons and ran out of gas. In one instance this world-class marathoner—who averaged under 5 minutes a mile during the 2014 Boston Marathon—shuffled through one especially rough mile in about 10 minutes!

"If I was bleeding, if I had fallen, obviously I would have completely stopped because I needed medical attention," he says of the experience. "But I was just depleted. So you can say, you know what? I'm done. I'm not going to go anywhere."

If you've ever had the experience of hitting the wall, you don't have to ask why it's called the wall. When it happens, it can literally—and I mean literally—feel like you have run face-first into a brick wall. It's palpable. Runners I have talked to have also described it as like trying to run with an elephant on your back, with cast-iron pots strapped to your legs, or while wading through wet cement poured up to your waist. As David Fisher, an Irish marathoner, says of one fateful race day when he hit the wall: "Fatigue is not the word, nor is pain. You need a whole other way to describe it. It's like a leech has sucked all the energy out of you, and your muscles have turned into rubber and want to contract."

Marathoner Beth Budnick Dembny says of her wall-kissing experience that she started to slow down after mile 21 or 22 "but was wishing for death at mile 23; mentally quitting crossed my mind, sadness that after 18 weeks of training and not missing a run, it wasn't enough to prepare me; pain in my legs was horrible, but even hobbling at that point was not going to happen."

Yokasta Schneider describes how at mile 22, "my fingertips and toes were tingling. I started crying and was so done and ready to quit. I didn't care about the fact that I had just gone 22 miles."

Okay, you get the picture. Hitting the wall is a multifaceted experience. It's a disheartening, painful race killer. But what causes this hit-the-bricks phenomenon? Is there any way to avoid it? Once you have hit it, is there any way to break through it? And what role does your brain play in all of this?

A PHYSICAL EXPERIENCE

The prevailing notion—and there have been a few over the years—has posited that hitting the wall is a purely physical phenomenon; as exercise scientists describe it, the result of a "peripheral limitation." As the theory goes, that overwhelming fatigued feeling and leg heaviness is the result of muscle failure; the muscles and liver have wrung every last drop of glycogen, the body's preferred source of fuel, from their stores. With no more gas in the tank, you are forced into a survival shuffle.

You've got about 90 minutes worth of glycogen stored in your muscles. When that runs dry, there's a bit more available for use in the blood. If you start your marathon at a reasonable pace, your fuel consumption ratio will be about 75 percent carbs to 25 percent fats. During the race, as your carb supplies begin to dwindle, the ratio shifts and your body begins to rely more heavily on fat for fuel. Once all your fat and glycogen blood stores are exhausted, the body is forced to throw the last log it has on the fire: protein. At the point you start chopping into your protein stores, the pain begins.

Many running experts believe that proper nutrition is the best way to avoid the wall. I'm sure you are familiar with the concept of carbohydrate loading, or "carb loading" as we athletes refer to it. There are lots of different carb-loading plans, but the gist of the practice is to pack as many carbs into your body as possible so you've got an extra supply stored in the blood and muscles to go the distance. Hence the tradition of eating pasta, pancakes, and rice a day and up to a week before a long race.

Even with a carb-loading plan, the most glycogen you can pack down for use during the race seems to be somewhere around 2,000 calories' worth. That's about enough to get you to around mile 16 to 20—exactly the point where most runners report bonking, though you might last a bit longer if you stick to a reasonable pace, stay hydrated, and maybe take in a few more carbs in the form of a sports drink or energy gel along the way. But be sure you have tested your formula well beforehand. On race day, nothing should be new or you risk hitting the wall due to digestion issues.

Of course, there are many other external factors that also influence how your body responds to a long run. Weather, training, sleep, medications, pace, family stress—these all make plus-or-minus contributions to whether or not you hit the wall.

THE CENTRAL GOVERNOR

Now for a more recent theory, and one that takes the brain's contribution to your performance into account.

Noted exercise physiologist Timothy Noakes doesn't argue that runners feel the wall physically, but neither does he consider it a purely physical phenomenon. He has come up with a concept called the "central governor model," which holds that the wall is actually a mental barrier. The central governor of the brain, Noakes hypothesizes, tells the body it's time to hit the wall whenever it feels the body has gone too far, too fast. When the brain determines you have reached what it considers your breaking point, it increases serotonin levels. This effectively reduces neural control to recruit muscle fibers, which in turn triggers the sensation of extreme fatigue. At the point you have a strong desire to stop, the negative thinking can easily ramp up into a chorus of negative voices (not hallucinations) telling you to stop.

According to Noakes's research, it's your brain that actually makes the decision as to just how thick that wall is. The brain chemistry of self-doubt dulls your mental edge to the point you become ineffective physiologically. In Noakes's view, this is just your brain's way of looking out for your flesh and blood; a way of preserving your health and putting a stop to an activity before you wind up injured. Your brain and body are like best friends who sync up to keep you from being "another brick in the wall." Pink Floyd's lyrics said we don't need no thought control, but I'm pretty sure that doesn't apply to runners.

If the central governor does exist somewhere in your brain, you might want to think of it as the voice of reason, the angel on your shoulder if you will, that's there to ensure you survive to run another day. Some might even say the central governor at times can be an overprotective guardian. I have had some runners tell me they have hit the wall as early as 3 miles into a race. These are runners who trained for longer distances, but for whatever reason their central governors were feeling particularly bossy that day and decided to shut them down in the early miles.

Although the governor may be whispering in your ear that you've given all you have to give, Noakes says in reality you may be able to dig deeper and give more physically. Certainly we all have seen a runner limping along the course who suddenly starts sprinting wildly when the finish line comes into view. Maybe you've been that person? It's as if the neurological pathways have parted like the Red Sea, allowing the physiological pathways to operate once again, providing just enough juice to kick to the finish.

By the way, we don't have to take Noakes's word alone as to the existence of the central governor. Some additional research in support of his theory exists. Most studies like this recruit cyclists instead of runners because it's easy to plunk someone on a stationary bike and measure their vitals as opposed to trailing after them as they jog down the road, but I think we can reasonably extrapolate the information to running.

A 2007 study[1] performed by researchers at the John Rankin Laboratory of Pulmonary Medicine found that exhaustion is indeed, to some extent, a mental condition. Three trials of cyclists working to varying degrees of perceived fatigue showed no differences in muscle fatigue even when performance varied. This seems to suggest the brain is the mastermind in charge of how tired muscles get, regardless of how hard they are really working.

Another study[2] had cyclists pedal away on a stationary bike while watching a relaxing documentary and then in another session spinning at about the same rate while performing a demanding mental task devised by the scientists. During the more laidback session, the athletes lasted 15 percent longer before hitting physical exhaustion, even though their heart rates and muscle work measured about the same in both sessions.

Then, in another test,[3] cyclists were asked to rinse their mouths out with either a sugary drink or a placebo drink containing no carbs at all as they rode a stationary bike to exhaustion. Those who swished around the real drink in their mouths outperformed the placebo group. To the researchers, this suggested the mere presence of carbs could power up the body even if the riders didn't actually have the benefit of knowingly ingesting them; their brains somehow recognized carbs were on the scene, and that seemed to be enough to provide an extra jolt of energy.

So is it your brain or your muscles that have the final say about when you have run far enough? Is there a central governor behind the curtain, pulling at the controls, or is it down to packing your muscles with as much fuel as possible and hoping for the best? It seems runners often pace themselves without really thinking about it. That shows there must be something in the subconscious at work, setting limits to how much energy you can use before going head to brick with the wall. And if the mind always has the final say, does that also account for the mind-over-matter mentality many athletes exhibit as they push past the point of endurance?

Personally, I think the answer lies somewhere in between. That mind-body connection you hear so much about is powerful stuff. I have seen evidence of it time and time again. I also think many serious runners seem to have the carb loading and physical side of things pretty well under control, but maybe haven't paid too much attention to preparing their brains to leap up and over the wall. While I do think eventually it's the physical side that has the final say, it's possible to delay and sometimes even prevent smacking up against the wall by having a good mental strategy.

If we assume the brain does indeed play some role in keeping your feet moving, it makes sense to persuade your gray matter to work in your favor. I've got some ideas on how you can do this.

CHOOSING A WALL-BUSTING MENTAL STRATEGY

To be clear, your brain is a pretty awesome organ, but it's no magician. If you didn't train properly or if you starved yourself for a week, you're asking to hit the wall no matter how great and powerful a mental organ you possess. But let's assume you behave like a responsible runner. In that case, I believe you can use your brain to influence your feet.

As you may recall from Chapter 6, there are four different brain strategies to choose from at any given time, each requiring a strong grasp of goal setting, visualization, and self-talk and each firing up different sections of the prefrontal cortex and the cerebellum. Let's do a quick refresher on those strategies so we can discuss which of them might serve as the best wall-avoidance technique:

- **INTERNAL ASSOCIATION** is a total focus on how the body feels while running; you tune in to the contraction and relaxation of your muscles, the mechanics of your arm movement, your breathing, your heart rate, and so on. Internal association's boundary is your skin and your focus stays inside it.
- With **INTERNAL DISSOCIATION** you do just the opposite, but still stay inside yourself. Using this strategy means distracting yourself by hitting

the mental replay button on a great song, writing an upcoming work presentation in your head, or counting your steps. No other runner around you even knows what you're thinking about. If they did, they may be surprised to learn that you weren't thinking about running.

- **EXTERNAL ASSOCIATION** places your focus outside your body and outside the act of running itself but keeps it on things important to your run. So with this strategy, you might pay attention to jockeying for position in the race, negotiating water stops, or your split times.
- Finally, **EXTERNAL DISSOCIATION** means focusing outward but on events or stimuli unimportant to the race itself. You may focus on the scenery flying by, cheering crowds, flower gardens in a park you pass, counting the number of times someone along the route yells the name written on your arm, or someone dressed in a weird running outfit.

Investigations into bonk psychology have looked at which brain strategies work best for the average, non-elite runner. One survey[4] of London Marathon finishers back in the '90s looked at how many average runners hit the wall and what they were thinking if and when they did. Internal dissociators (those who used distracting thoughts) were the most likely to hit the wall, but when internal associators (those who tuned in to bodily sensations) hit the wall they did so earlier in the race. External dissociation (focusing on scenery, cheering crowds, etc.) seemed the most effective wall-avoidance strategy of all and resulted in a later onset of fatigue.

On the face of it, it seems counterintuitive to avoid inwardly directed thoughts. Monitoring how your body is feeling would seem essential. According to the research, most elite runners seem to benefit from internally directed thoughts—just the opposite of us mere mortals. That could be because they're better at dealing with the information they receive about their bodies and are better equipped to deal with any bad news. We have to be careful not to apply everything that works for a world-class runner to the average runner. Although we can learn a lot from how they train and think, not everything they do will apply to the rest of us. It's worth noting, too, that the London Marathon researchers did not look at the elites in their work.

So, looking at ordinary runners, the Brit scientists found that too much internal focus, whether it was related to tuning in or tuning out, appears to set up a negative conversation between two brain regions, the anterior cingulate cortex (ACC) and the insular cortex, also known as the insula.

The ACC's job is to notice anything important, and the insula's job is to sense pain. Of course, you want to tune in to anything that could be serious trouble. But if the insula goes into hyper-drive and starts griping about every little ache and pain and the ACC becomes a bit too sympathetic, the two brain regions begin to dissect even the slightest discomfort ad nauseam. Worse yet, they sometimes ask for the opinion of the highly emotional and super-reactive amygdala.

Internal dissociation is actually the worst of all thinking strategies if you are trying to avoid the wall. Distracting yourself from the body's sensory signals and other important aspects of running can cause you to misread the situation. Because you're so busy internalizing thoughts unrelated to anything essential to your run, you may misjudge your pace or forget to hydrate properly or fail to notice some other vital body signal such as pain.

So what's left? The two external strategies. Interestingly, runners who stay externally focused, particularly in a dissociative manner, don't seem to hit the wall as hard or as often. Perhaps directing your thinking outward keeps the ACC and the insula from gossiping too much and prevents the amygdala from entering the conversation altogether. A cheering crowd, a banner hanging out of a window, or a band playing in the distance may be just enough to distract your brain from the punishing bodily sensations of running without causing you to lose too much focus on pace, water stops, and the like.

I suspect it may be ideal to do periodic but brief check-ins with internal factors, because your body was designed to use pain as an alert system when something is off. Ignoring pain at all costs is never a good idea. Take it from me: I have seen when people disregarded pain and suffered the physical and emotional consequences.

My colleague and friend Dr. Matthew Buman, an assistant professor of exercise science and health promotion at Arizona State University, has conducted his own research into runners and the wall. Interestingly, he

WRAP-UP

I believe in the central governor model, but I think there's more to the story. Just as your brain tries to protect your body from harm, I think perhaps your body tries to return the favor. Your brain uses carbs for fuel, too, and a lot of them. Once you start running low during a race, you not only feel tired, you may feel depressed, discouraged, and irritable as well. Perhaps your muscles start to shut down as a way of preserving fuel for the brain. Don't be surprised if you start feeling that way. Honestly, it may feel a bit like depression because serotonin has been depleted. Difficulty concentrating, poor motivation, sluggishness, negative thinking, and hopelessness all can become part of the picture at that point. Your job is to know your brain, how it works, and how to respond when things start to shift from positive to negative.

I also believe this: No matter what theories are true or what strategies you employ—physical, mental, or otherwise—there is no surefire way to avoid crashing into the wall face-first. The protective mechanisms of both brain and body are there for a reason. Maybe you can stretch your limits to some extent, but it's always a possibility that fatigue will get the better of you.

Buman's research suggests that self-talk and self-expectations play a huge part in wall avoidance. That's why I emphasize doing a visual-

has found that men are more likely to hit the wall than women. His research also suggests that expectations play a big part: If you think you might hit the wall, you are more likely to do so.

ization of hitting the wall and seeing yourself deal with it effectively. If you believe you are going to dominate the wall, you are more likely to see your beliefs become reality.

If you do hit the wall, please remember it can affect your ability to think. I have seen plenty of people like Steve Crossland, disoriented and slurring their words, hustled into the medical tent. I feel concern when I see them out on the course wobbling through runner traffic, trying to make it to the finish. In 2014, a woman in Boston ran across the finish line, ran into my arms, and just kept running. She couldn't stop. Her legs were not communicating with her brain any longer. There's a point of no return you need to accept. Going beyond that can be dangerous. No race is worth compromising your health or safety.

So be aware that if you do hit the wall, it might not be just your quads that are in question. You may not be thinking clearly either. For this reason, it's a good idea to ask for help. Have someone keep an eye on you as you recover, especially if you choose to keep plodding through to the end. Try to find someone to coach you through it. Sip on a sports drink to get some carbs into your system, but don't overdo it. If it's not your day, it's not your day. Better to keep yourself safe and pack it in if you truly can't go on. There's always next time.

CHAPTER 15

Running Through Challenges

Toughing it out when things are rough

THE IDEA OF struggling and suffering and bleeding from the toenails may sound like a fun adventure exactly never, but when I talk to my marathoners and ultramarathoners they seem to live for that kind of pain. For that matter, so do sprinters. And so do middle-distance runners.

The fact is, runners as a group are generally as tough as bunions. Whether they tune in to or tune out of their pain as a main brain strategy, so many of them embrace the twinges and dings that come with the sport like a badge of honor. No one wants to hurt so badly they're out of the game, but for a lot of us, we haven't run if we don't have a body part to soak or bandage.

Still, every runner I have ever met has a kryptonite. I bet you're no different. What other runners take in stride might feel to you like the last hurdle of a steeplechase. I always say that it's not the body that throws the greatest challenges at you—it's the mind. Just as you might have a tight iliotibial (IT) band or a tweaky knee to overcome, you probably have a few running psych-outs that challenge you as well.

What is it that takes you out of your comfort zone? Is it something you choose as part of your training, like high-intensity work? Or is it something beyond your control, like a hill that drives you bonkers? Let's use this chapter to talk through five big challenges so many runners seem to come up against, and find a way to break through. (I use the next chapter to discuss the trials and travails of weather—they are a special kind of challenge in their own right.)

Whatever it is that challenges you (and sometimes it's more than one thing), I will remind you about some strategies you have already learned from previous chapters. Plus I will enlighten you with some insider tips I have borrowed from my runners. No doubt, some circumstances may remain more difficult for you than others. But the next time you come up against your personal Waterloo, at least you'll have some tools to work with.

HILLS

When you say things are looking up to a runner, don't be surprised if he views that as a downer.

To many a runner, "up" means hills. And hills are the enemy. They are an obstacle standing in the way of fast times. A burden to be endured. A soul-sapping, lung-bursting exercise in pain.

Remember, I'm the psychologist for the Boston Marathon, home to Heartbreak Hill, one of the most feared stretches of incline in the world. Over the years, I have watched people of all abilities face Heartbreak with all sorts of emotions and outcomes. I have seen fear, anguish, pain, and rage. But I have also seen confidence.

I've had the opportunity to quiz runners of all ages, levels, and abilities about how they approach the ascension of Heartbreak. The ones who conquer it without stress have given me a few good ideas to share—and you can find even more in Chapter 13.

Hill Strategy #1: **Love Thy Enemy**

So many times a runner will come to a hill with a preconception of how horrible it will feel to run up it. Those negative feelings form a feedback loop in the brain, stoking your hatred of hills even more. When you come to the bottom of the hill with thoughts like that in your head, you set yourself up for a miserable experience.

What you really should be doing is convincing yourself how much you love those hills. Tell yourself that hills are the greatest thing ever, rather than something to be feared. After a while, you will believe.

You can also focus with some good old-fashioned positive self-talk. Tell yourself you're the little engine that could, that slow and steady wins the race, that what goes up must come down—whatever cliché helps you embrace the feeling of chugging up, up, up. Look at it this way: The hill is not going anywhere, so you may as well make the best of it.

To reinforce your hill-loving ways, incorporate regular hill work into your training for a few weeks, and make mental training as much a priority as physical training. A nice bonus here is that practice makes perfect: Running up the hills will make you a better hill runner in general, which in turn will make you a more confident hill runner.

Hill Strategy #2: **Imagine It's Easier**

Come up with imagery that helps you get up the hills. One runner told me she sights something along the edge of the road, such as a tree or a car, then throws a mental rope around it that she imagines she can use to pull herself upward. Another runner told me he pretends he is being carried up the hill by a winged horse. Someone else told me he imagines a large truck behind him pushing him upward. Marathoner and coach Jeff Galloway calls these "dirty tricks." These are the imagery, words, and thought loops you keep on the back burner until you need them. You use them to get past a problem and then discard them from your thoughts.

Hill Strategy #3: **Tune Out or Tune In**

Some runners go to their happy place to try to forget they are working so hard. Or they dissociate by, say, pretending their legs have become detached so that the burning, aching lactic acid sensation in their quads and hams no longer belongs to their body. Other runners I have spoken with take the exact opposite tack: They own their pain with a sort of "hurts so good" attitude. Cramps and feelings of fatigue only make them push harder. Experiment to find out what works for you. Flip back to Chapters 6 and 7 if you need some additional ideas on how to best use association and dissociation.

FEAR OF FAILURE

When Jack Schneider was relatively new to running, he toyed with the idea of running a race. What held him back was speed training. He was all about the time commitment and totally willing to put in the miles, but he found the idea of speed work quite intimidating.

A lot of runners are afraid to push the pace because they're scared they may not be up to the task. I have known just as many folks who fear longer distances for the same reason; they don't believe they will ever have the wherewithal to go the distance.

Every runner has moments of doubt. This is not entirely a bad thing. It reflects your investment in your efforts: self-intimidation can be self-defeating and prevent you from moving forward. These feelings are rooted in negativity and self-doubt. They can become a self-fulfilling prophecy that affects you physically and distracts you from your goals. Here's how to get past common sources of intimidation and run your best, without doubt.

Self-Doubt Strategy #1: **Feed Your RAS**

Your reticular activating system, or RAS, is the part of your brain that holds on to your beliefs. One of its most important jobs is to scoop up any information that either supports or refutes your view of yourself to store for future use. Both self-doubt and confidence flow from your RAS. You can control what pours from this area of your brain and into your thoughts by filling up the RAS's tank with information, ideas, thoughts, and feelings that support success. Flip back to Chapter 3 for plenty of ideas on how to do that. You will also find a positive thinking worksheet in Chapter 18.

Self-Doubt Strategy #2: **Know the Warning Signs**

Anxiety, doubt, fear: all perfectly natural feelings, and perhaps even helpful to a point. There are circumstances where a healthy dose of fear can be useful and even wise. Before the start of a race, for example, a little bit of jitters means your adrenaline is flowing. But dwelling on negative feelings or ignoring them can hold you back.

Watch for any sign your fears are impeding your progress. If so, practice some mental strategies to conquer your doubts. Definitely be thinking in positive affirmations. And use visualization to picture yourself overcoming whatever has got you intimidated. For example, you don't want to look bad having so many runners streak past you in the final kick of a race, so use the last few minutes of every run to pick up the pace. This will teach you that you can indeed run faster, at least in short bursts. Maybe next time not so many runners will pass you—and perhaps you will even pass a few people yourself.

Self-Doubt Strategy #3: **Break the Ice**

If you're afraid of doing something, just do it. What's the worst that can happen? There's a chance you are not as bad off as you think. If you are, you have a baseline to work from, and with training you will get better.

Take baby steps. If like Schneider you think of yourself as a permanent slow poke, you might start incorporating one or two short speed intervals into an otherwise ordinary run. You will get a slight taste of what running faster feels like. That's something you can gradually build on. Perhaps you will never be Usain Bolt, but who is? You have to set your sights on something that is realistic and achievable.

Self-Doubt Strategy #4: **Set Up Success**

If hitting a certain time for a 5K or being able to complete a certain distance is super intimidating, focus on improvements rather than total outcomes. It all comes down to those best/great/good goals. You may not be able to turn in a PR in your next race, but maybe you can shave 10 seconds off your previous time. Or you're aiming to complete a 10-mile distance but you only make it to mile eight—if it's a half-mile more than last time, you have to give yourself credit for doing better. Use your training time to work on building confidence, both physically and emotionally; train your body and your brain for the goals you want.

BOREDOM

There is one running experience that has become synonymous with boredom. It's called treadmill running. It's such a mind killer that I have devoted an entire chapter to it (see Chapter 17). Perhaps you may decide to skip that chapter because the only thing that sounds more boring than running on a treadmill is reading about it. Trust me, though, there's information there that can help you when you're faced with the choice of running on a treadmill or skipping a workout.

You might think that so long as you're outside your mind will never go numb, but that's not the case for everyone or for every run. While running outdoors can be exciting, challenging, and inspiring, it can also be dull and repetitive. Your brain desires novel experiences, which the reward centers of the brain respond to by releasing dopamine, a neurotransmitter that brings on sensations of joy and delight. It really doesn't matter if you are on a treadmill, running through the streets of Paris, or skimming along the Forest Park trails in Portland, Oregon—once the novelty of running wears off, the faucet for those feel-good chemicals shuts down and you may as well be listening to a lecture on the organizational management of office cubicles. That's the organic chemistry of boredom in a nutshell.

Boredom-Busting Strategy #1: **Distract and Dissociate**

There are definite advantages to associative strategies that force you to pay attention to things like stride, form, and posture. Association can make you a better runner. But if you find yourself in a phase where boredom is an issue, dissociation may be the way to go for the majority of runners.

By all means, put on those headphones and crank up the tunes (so long as you can do this safely) if that's what it takes to keep boredom at bay. Put together a motivating playlist that engages your brain and takes your mind off the road.

Char Beasley tells me she got a serious case of the doldrums one winter. Usually a loner, she decided to pick up a running buddy. "The miles go by a lot quicker now," she says. "Sometimes we talk, but even when we don't I think there's something to having someone alongside you that makes it a lot more interesting."

Sammy Chang found a creative way to make it interesting. As he runs through the streets of New York City, he looks for loose change—pennies, nickels, dimes, and quarters. When he spots a coin he scoops it up and puts it in his pocket. At the end of the month he tallies up his spoils, and if he has enough he buys himself a tune, an app, or the like. Amby Burfoot swears this was a favorite activity of his running partner Bill Rodgers, too!

"Most of the time I don't clear more than 50 cents in the month or so, but I've hit five bucks a couple of times and once I even found a 20!" Chang says. He has also picked up a few interesting non-monetary items along the way, and seen a lot of pretty gross stuff, too. Obviously, his strange hobby isn't making him a millionaire, but he says it keeps him eager to hit the road.

Believe it or not, you should also consider the treadmill. Yes, the "dreadmill," which you and everyone else on the planet associate with boredom, and to which I have devoted an entire chapter. Park yourself in front of a TV screen and laugh your way through a comedy or lose yourself in a great movie. You may not necessarily run your best or your prettiest, but you will be moving. If you only step onto a treadmill occasionally, it might actually wind up being a refreshing change. (Then again, it might not: This tip definitely won't work if you're one of those people who truly despises running on a mill.)

Boredom-Busting Strategy #2: **Change It Up**

Most of us tend to autopilot through one, two, or perhaps three different routes. Plus, you probably run around the same time every day and at the same pace. A few years, months, or weeks of that—depending on your tolerance—can start to feel pretty rote. I have one friend who runs at lunch time around an office complex. There aren't too many places I can think of more boring to run than that.

If you always do your bridge out-and-back, head to the lake, the track, or the park instead. Toss in some fartleks or intervals, or change it up completely by doing hill repeats or speed ladders. You may want to get away from running entirely and toss in a few days a week of cross-training. To fire up those novelty-seeking brain chemicals and get the RAS paying attention, create novelty.

Boredom-Busting Strategy #3: **Break It Up**

When I've got a runner who is in a really unmotivated phase, I have them split their runs into two or three daily sessions. This is counterintuitive, yet it often works.

Sometimes the idea of three 20-minutes hits of exercise is easier to face than a continuous 60 minutes. You don't give yourself enough time to get bored during such a short stint because it's over before you get started. You still aim for doing your full mileage, which means keeping two or three appointments a day, but that doesn't seem to be a barrier.

This is a particularly useful strategy if you get bored on long runs. You can still get the mileage you need, but because you break up the run in bite-size bits, it becomes more palatable.

LIFE CHANGES

You will go through many phases in your life that may change your relationship with running either temporarily or permanently. It's worth reviewing some big phases you will likely go through and provide a tip or two on how to carry your love of running with you through life. I'm calling out specific life events here, but many of the tips are interchangeable for just about every major shift of circumstances.

Life Change #1: **Becoming a Parent**

Joan Benoit Samuelson tells me that she refers to her running career in two phases: BC and AD.

"That's before children and after diapers," she explains flatly.

After Samuelson's kids were born, she cut way back on mileage and went from running twice a day most days to running just once a day. This left her feeling anxious and out of sorts. As is the case with many new parents, she learned that there are few things that change your life so drastically as having a baby. Lack of sleep alone is enough to throw anyone's life into turmoil.

If you're the mom, you already have a taste of things to come after nine months of pregnancy. You have experienced exhaustion, nausea, and a morphing body. Although you hear about the occasional mom-to-be who runs marathons (and sometimes wins), you likely came to your own conclusion that pregnancy is generally not the time to be shooting for personal bests.

During pregnancy, moms need to take it one day at a time. Once again, goals are great, but so is perspective. If you don't feel up to a run, don't push it. Forcing yourself through a workout won't help you, and it's likely not ideal for the baby. I know competitive women who have had to sit out most of their pregnancy on doctor's orders or because their bodies couldn't handle it. Taking a break to have a baby doesn't mean you're not tough or dedicated. It means your child comes first.

And you must listen to your body. Alexa Simpson ran pretty comfortably during most of her pregnancy. But right as she was entering her ninth month, she stepped off a curb one day and felt the ligaments in her knee go ping. The hormones that soften the joints during pregnancy finally caught up with her and she knew it. Sensibly, she realized enough was enough and slowed down to a walk for the rest of her pregnancy.

"You don't want to ignore that kind of warning," she says. Once the baby is born, I tell new moms to give themselves a break. Their bodies have been through a momentous event. You can't expect to jump back to the same times and distances you were doing before you carried a baby in your belly. Allow yourself 6 to 12 months of slack before setting tougher goals and expecting more from your body.

By the way, because babies change your priorities, schedules, and time commitments, I tell new dads the same thing.

I also advise new parents to run completely different running routes and do different workouts after the baby is born in order to avoid comparisons pre- and post-baby. Mentally, it's a bummer when you don't feel you measure up to your usual standards. Rather than beating yourself up, be kind to your body with activities like stretching, massage, or yoga.

Postpartum appearance is another way new moms (and sometimes new dads, too) come down hard on themselves. I have known more than a few new moms who came home from the hospital and immediately tried to squeeze into a pair of pre-pregnancy pants. After doing one of the most amazing things the human body can do by giving birth, they didn't even allow themselves a few days of "mommy tummy." Now that's being hard on yourself! Rather than hanging your self-esteem on outside appearances, purchase a couple articles of clothing including a few running outfits that fit you right now, and give yourself a break.

Samuelson suggests finding ways to adapt so you can continue to hold on to some of your old ways of training until you get more time. She tells me she never medicated for her anxiety.

"I think it was because my endorphins weren't elevating like they were accustomed to. I finally decided to go out and do one long run per week that would sort of deplete me. Within 2 or 3 weeks of running one 13- to 15-mile run a week after dropping down to 10 or less—that took care of the anxiety almost overnight.

"Because of having gone through that, I decided I needed to teach myself how to ride a bicycle so that if the day comes I can't go out and run, at least I can expend my energy doing something else. When you are not able to expend the energy you are accustomed to spending, you build up a lot of adrenaline in your system and it has no place to go. That's my hypothesis. It's not documented with any research, but I've been at it long enough to figure out some things pretty simply."

Life Change #2: **Getting Older**

If you were a decent high school or college runner but you're long past graduation, your times may be slipping. That can sting. A lot. Those added minutes and seconds on the clock are a reminder that, try as you might to fight it, time inevitably marches forward.

"I may have the world's record for slowing down the most over my career," former elite runner Amby Burfoot told me. "I won the Boston Marathon when I was 21 and ran my best marathon when I was 22 and I've done nothing since then, and I've gotten slower through the years."

Burfoot is probably selling himself short. He is still pretty active and still finishes up decently in his age group in races. But he does have a point when he says you may have to rethink training and goals after a certain number of birthdays. Maybe you used to do five big marathons a year. As your body ages, you might pick your spots and do only three. That's a psychological adjustment, but it is probably essential for staying healthy and injury free.

It's always important, but even more so when you're aging, to seek out running peers. You want a reality check and a bit of commiseration. Rather than comparing yourself to the latest teen sensation, age peers help give you a more accurate comparison to where you are with running. That's why I try to get runners who compete to look at where they are within their age groups. You may not be able to keep up with the twenty-somethings any longer, but if you can stack up somewhere in your age group, you know you're doing okay. It's an accurate form of victory.

Even if you do slow down, don't retire from running. You are never too old. Octogenarian Harriette Thompson recently smoked the competition at the Rock 'n' Roll Marathon in Washington DC, smashing the record in the 90–94 women's age category by more than an hour and a half. A day after the race, Thompson told me her legs were recovering pretty well. Considering she had recently endured nine rounds of radiation on them for cancer treatment, that's pretty darned impressive. Plus she raised more than $90,000 for the Leukemia and Lymphoma Society's Team in Training.

"I hope to run the race next year even faster," she told me. "If I'm still around."

What a gal. She didn't even start racing competitively until the age of 79. She's not the only one either. I have spoken with several senior racers, including one man from India who claims to be 101 years old. They always tell me they keep their age in perspective. Maybe it slows them down a little, but it never seems to stop them.

You should still have goals, but as Burfoot wisely says: "When I was younger I wanted to be fast and to win, but now I have longevity goals."

"I think there are so many reasons to pursue a life of high fitness through running," he continues. "I don't care what yours is so long as you find one and latch onto it, and use it as a friend to keep you out there. I am 68 now—I know I am going to die some day, and in the meantime I would rather be as fit and active as I can be within reason, rather than sitting on the sidelines watching the last quarter of my life go past me."

As the late Dr. Kenneth Cooper used to say: "We do not stop exercising because we grow old; we grow old because we stop exercising!"

Life Change #3:
A Major Stressor

This can be bad stress: illness, divorce, the death of a loved one—sometimes all at once. Or this can be good stress: marriage, a growing family, buying a new house. But to your mind and body, stress is stress. Your body really can't tell the difference between distress (negative stress) and eustress (positive stress). Someone building a new house could be susceptible to the same range of physical and psychological symptoms as someone whose house has just burned down.

When something stresses you out, your body reacts, thanks to an overwhelming response from your sympathetic nervous system. For starters, your adrenal glands release a flood of the stress hormones adrenaline and cortisol into the bloodstream, which elevates a host of involuntary functions including heart rate, breathing, and blood pressure.

Lots of other things happen in the brain and body, too. Your blood vessels open wider to allow more bloodflow to large muscle groups, just in case they're needed to flee the scene. Your liver releases some of its stored glucose for a shot of instant energy. Your pupils dilate to sharpen vision, and your sweat glands produce sweat to cool the body.

A little burst of nervous energy every now and then can boost your ability. The problem is that repeated, prolonged stress can wear you down, causing a laundry list of physical and emotional problems including everything from headaches to hair loss to weight gain. The American Institute of Stress publishes a list of the 50 most common symptoms—50! And those are just the most common. I think there are just as many reactions to stress as there are people.

The first thing someone struggling through an exceptionally bad or good patch often lets fall by the wayside is his exercise routine. But that's like tossing out the medicine that can cure you when you're sick.

Running is by far one of the best ways to help manage your stress. Exercise can help reverse the damage brought on by stress by pumping out neurohormones that help keep the brain healthy. It ensures your body stays healthy, and as I like to say, anything that's good for the body is doubly good for the brain. But you probably know how good running is for you already. The challenge is to keep at it when you already feel overwhelmed.

I know it can be difficult to get moving when you have a million things to do and a million thoughts racing through your head, but that is exactly what you should do. The drill is to set aside time every day in your calendar and keep that appointment like it was a meeting with your boss or having a doctor's appointment. Easier said than done sometimes, I know, but try.

One trick I find works well is to cut your workouts in half. Do shorter, more high-intensity workouts. You'll get done faster, plus high-intensity seems to have a calming effect on many people. This is because it clears cortisol and adrenaline from the system pretty quickly. It's like the modern-day equivalent of sprinting away from a saber-toothed tiger.

Life Change #4: **Injury or Illness**

Injuries can be just as much a mental challenge as they are a physical one. Part of your identity is etched into the RAS and other parts of your brain, so not being able to run leaves a void. Sidelined runners often don't describe an injury in terms of aches or pains; they talk about feelings of depression and anxiety. As the great marathoner Bill Rodgers says: "There's nothing like having something taken away from you to make you realize how much you love it."

Instead of moping around waiting for your body to repair itself, use your downtime productively and positively. Here again is an opportunity to reframe a problem as an opportunity. When you look at it that way, injury can be a great motivator. Runners tend to thrive on routine, so try to stick to the other aspects of your life that revolve around your daily run such as sleep times and eating patterns. And use the blank space in your day to find new outlets for your energy. For example, take up weight training, Pilates, or yoga; you know you've been meaning to find the time to work on your strength and flexibility anyway.

Sharon Jackson, who blew out her knee toward the end of a grueling road-racing season, tells me she used her 6-week layoff to work on touching her toes, something she never was able to do before. "I'm a typical runner, very tight, which probably contributed to my injury in the first place," she says, adding that once she was cleared to run again the extra stretching really paid off in the form of a looser, longer stride.

If you have had your wings completely clipped, try something nonphysical that can help your running, like meditating to improve your brain skills. Or refocus your energy entirely. Why not tackle that project around the house you've been meaning to get to? The point is to avoid spending your entire downtime on the couch and feeling sorry for yourself. Provided you don't aggravate your injury, you can distract yourself.

To keep the feeling of connection, stay social with other runners. You can surf through running communities on the Internet or see if you can entice some of your running pals to join you in one of your cross-training endeavors. I encourage you to show up at a few races to cheer your friends on if you can. You might think this would stir up pangs of jealousy, but I find more often than not, supporting your friends makes you feel good.

Above all, if you get injured, put on the brakes. "Hurts so good" may work on a hill, but it's not the way to go here unless you want to risk being permanently sidelined. Although it may be a few weeks or maybe months before you're able to resume normal activity, appreciate the improvements you make on the road to full recovery. Think of goal setting in this circumstance as your way back from injury.

I like the way ultra runner Dean Karnazes puts it. As someone who has run for 3 days straight without sleep, he has known his share of injuries. To get through them, he tells me he is all about keeping his goals in sight.

"Goals to me are short-term milestones and are an endless procession of ongoing 'baby steps' on the path to reaching a dream," he tells me. "For me, I start with a dream and work backwards in aligning goals that will help get me there."

WRAP-UP

Running can be like a deep relationship. You may love it but not love everything about it. Yet just as you accept the flaws of the people you're fond of, so it goes with running. There will always be hills to climb, boredom to overcome, or injuries that dog you. The secret to running longevity is finding a way to live with the aspects of it you would prefer to live without. As this chapter has shown you, it can be done.

CHAPTER 16

Weathering a Storm

Psych yourself up for any kind of weather

ULTRA RUNNER MORGAN IVES is especially masochistic. While mere mortals succumb to their lactic acid threshold within hours, Ives often runs for an entire day without stopping. He wracks up minutes, hours, miles, days, always with a smile on his face and nary a muscle ache.

But in the same way some people prefer not to be in the same room as a spider, Ives has an aversion to running in the rain. Even a light rain. Even a sprinkle.

"Just don't like to get wet," he tells me simply as he is literally globe-trotting through an airport on his way to a race called the East 24, a 24-hour ultra in the Midlands of England. He has nothing else to add. He hates running in the rain. That's all there is to it.

While running ultras may seem superhuman, Ives is indeed still human. Neither rain nor snow nor sleet nor hail stops the mail carrier. But Ives? A few raindrops is all it takes.

In the previous chapter we talked about some aspects of running that runners struggle with. For whatever reason weather presents its own special trials for a lot of runners. Once I started thinking about the subject, I realized I had quite a lot to share with you. I realized that weather, like your significant other, is not always perfect. And if you wait for climate perfection, you will miss a lot of days. You can get to your "make it work" moment by adjusting your attitude and hitting your stride no matter what Mother Nature is dishing outside.

HEAT

Let's start with running in the heat. I think hot weather stops a lot of runners in their tracks. I was surprised to learn from veteran runners and Boston Marathon winners Kathrine Switzer and Jeff Galloway that they aren't so thrilled with running in the heat and sometimes skip or edit workouts because of it.

Some runners, like Jen Mahoney, have a genuine aversion to high temperatures. While this might sound melodramatic, Mahoney tells me she views running in 90-degree weather as a metaphysical duel with the sun. She lives in Arizona, so this duel occurs more often than she would like.

"Just last week I found myself lying in bed the night before what I knew would be a scorching day, all stressed out," she told me, adding that the thought of everything from sunstroke to sunburn was racing through her head.

Mahoney got herself up and out by 6 a.m., but by then it was close to 86 degrees and the sun was already baking the pavement. "I ran for maybe two blocks and then bailed," she says. "That's pretty much how it went all summer."

To a point, Mahoney was right to be concerned. Running in the heat can lead to heat exhaustion or worse. Sun-related conditions can cause both physical and mental problems. Many a runner has wandered into one of the medical tents at Boston confused and delirious as a result of too much time in the sun.

Dehydration is of course always a worry, and can cause both physical and mental disruptions. Another side effect of running in the heat is a condition known as hyponatremia, which is sort of like the opposite of dehydration. When a person drinks too much fluid, it literally causes a flood within the body that dilutes the concentration of sodium in the blood. Cells are forced to absorb the excess water, which leads to potentially fatal swelling in the brain. Among other things, it can trigger mental confusion.

I mention hyponatremia specifically because it brings to mind my first year as the official psychologist on the Boston Marathon medical team in 2002, when a woman came staggering into the tent claiming she had lost her sight. After a series of tests by various physicians, we all put our heads together and came up with a hyponatremia diagnosis.

Novice female marathon runners who clock in times over 4 hours are often the most susceptible to this condition. They tend to naively and compulsively guzzle water throughout the race to stave off dehydration, just as they have been told to do by countless experts and magazine articles. When this sound advice backfires and they overdo it, the victims develop salt-caked skin, a slight increase in body weight as they cease to perspire or urinate, and very off-center behavior. I find it to be a classic demonstration of the powerful brain-body connection I have observed working with runners over the past 12 years as a clinical sport psychologist.

So yes, running in the heat is no joke. Even if you're good at it, take it seriously.

When you must face the heat—perhaps because you are racing or have a group run planned—there are some mental strategies that can help you through.

Heat Strategy #1: **Adjust Your Expectations**

As much as I push goal setting, a scorching-hot day may not be the time to reach for the stars. Sometimes mentally recalibrating is not the easiest thing to do, especially if you are in a race situation or trying for a qualifying time, but that's why I always tell my runners to have a set of best/great/good goals. Maybe the heat prevents you from reaching your best goal, but you can shift your goals into great or good mode and still feel successful.

Heat Strategy #2: **Create a Cool Loop**

Mahoney and I came up with one system that can potentially work very well for you. I had her set up a short loop around her neighborhood where she stows a cooler stocked with cold drinks, ice, and a towel in the shadiest spot she could find. She is able to stop every 10 minutes or so for a cooling break. Psychologically, this breaks up the distance and gives her something to look forward to, sort of like an aid station at a race. Of course, this helps manage the physical symptoms of heat running, too. After a week you can gradually lengthen your cooling loop to stretch out the time between stops.

Heat Strategy #3: **Know Your Danger Zone**

Despite your best efforts to run on the edges of the day, drink plenty of fluids, and dress appropriately; a hot day always means there's a risk of developing a heat-related illness. Know the signs and symptoms. The physical ones are often obvious—dehydration, a rash, fatigue—but do you know the mental signs?

Disorientation and confusion are common among people with heat-related illness. If you start to feel out of it, stop and seek help. Use your cellphone to call 911. If you can't find anyone to help you, at least try to get to a shady spot and cool off and take a drink. And be a good runner-citizen by watching for heat problems in others.

COLD

For cold weather you can apply some of the same tips you use for hot weather, and many of the hot-weather psych-up tips in reverse. It's unlikely you need to worry about hyponatremia, but dehydration—and all the accompanying physical and mental issues that go along with it—is still a worry.

Cold Strategy #1: **Respect Your Limits**

The main thing you need to do is adjust your expectations. You're probably not going to run a PR if you're slipping and sliding down an icy hill or shouldering a windy snowstorm. Marathoner Laura Oldenburger tells me she did just that while running a race in upstate New York one February.

"It was me and, like, six other runners when a snowstorm suddenly rolled in at around the 18-mile mark. We couldn't see the mile markers or the orange cones or any of the race officials, so we wound up taking a wrong turn," she recalls.

By the time anyone figured out they were missing, the pack had veered nearly 3 miles off course. When someone finally tracked them down, a few of the runners decided to call it a day. Not Oldenburger. She decided to course-correct and finish the race. She wound up limping across the finish line with 6 extra miles on her legs. And she also came in second place.

"Best finish ever," she told me with a smile.

Oldenburger certainly adjusted her expectations that day. It was her worst marathon time (by more than an hour and a half!), but she was pretty pumped about coming home with a trophy.

Cold Strategy #2: **Snow What You're Doing**

When your body temperature drops to the point you need to shiver to keep warm, you're in the danger zone for cold-weather injury. Your bloodflow to the brain slows, and you can feel sluggish and confused.

The best defense against these symptoms is to avoid them in the first place. On a blizzardy, snowy, icy day, staying indoors is best. That's not a cop-out. It's common sense.

But if you find yourself in a dangerous situation, seek shelter and help immediately. (Oldenburger was probably pushing the limits. If the weather reached blizzard conditions, my hope is race officials would pull runners off the course.) If your clothing is wet, remove it. Reheat yourself with blankets and warm rather than scalding fluids. Hypothermia is serious business, so seek medical attention as soon as possible.

PRECIPITATION

That old saying about neither rain nor snow nor sleet nor blah-blah-blah doesn't always apply to runners. Some of you hate to get wet. But don't worry—I promise you, you won't melt if a raindrop or snowflake touches your skin. Like I tell runners in New England: "If you don't like the weather, wait a couple days and you'll hate it even more!"

Precipitation Strategy #1: **Manage the Moment**

Now if you're like Ives and inclement weather gets you down—rain, snow, sleet, whatever—you can try and tune in to it and make yourself focus; but honestly, I would go with a strategy that's dissociative but not too dissociative. Though I am not a fan of cranking the tunes if there is low visibility and less of a chance you will be spotted by oncoming cars, you can certainly run with a friend to help take your mind off the misery. Find a balance between ignoring the downpour and protecting yourself from danger.

Precipitation Strategy #2: **Reframe the Negatives**

Running in the rain stinks? No, running in the rain is great! It's an awesome challenge! It's a cool experience! As they say, you've got to fake it till you feel it. Once you get going, you almost always feel better. You learn to accept the fact that you are going to get wet and that you will survive. Getting yourself out the door and taking that first soggy step is the hardest part. After that, all the negativity rolls off your back along with the rain.

WRAP-UP

"Climate is what we expect. Weather is what we get," Mark Twain once said. In other words, you live in a general climate, but more often than not you will have to deal with weather.

The more you respect scorching heat, bracing cold, and whatever form of water falls from the sky, the less likely it is to interfere with your running program. That's not to say it's a good idea to take a jog in a tornado. But if you come prepared with the right gear, the best information, and a great attitude, weather doesn't have to be the enemy.

CHAPTER 17

Treadmill Training

Race to nowhere

IN 1822, THE AMERICAN PRESS began praising a new invention from England that struck fear in the hearts of convicted felons everywhere. Several American prisons quickly adopted the idea, hoping it would match up with the glowing reports they were hearing from prison officials across the pond.

It did, and it didn't. Although some administrators praised the device for keeping prisoners orderly and submissive while requiring little supervision, others noted that it didn't exactly inspire a transformative rehabilitation among the incarcerated.

"The labour of the tread-mill is irksome, dull, monotonous, and disgusting to the last degree," wrote a British prison official at the time. "[T]he treader does nothing but tread; he sees no change of objects, admires no new relations of parts, imparts no new qualities to matter, and gives it no new arrangements and positions."[1]

So now you know: The treadmill, often billed as the gold standard for weight loss tools and improving cardiovascular health, was originally used as a form of corporal punishment. Perhaps you already sensed this during a time when you were trudging along on a treadmill, belt slipping by beneath your feet, the timer on the console agonizingly ticking off the seconds, your inner voice begging you to punch the stop button.

I bet there isn't a runner alive who doesn't loathe the torture of grinding out a workout on the treadmill, am I right? Even with state-of-the-art entertainment to help you pass the time, it seems virtually impossible to distract yourself from the fact that you're going nowhere, and probably not fast enough.

Your brain does not appreciate boredom. During novel and entertaining experiences, the reward centers of your brain freely dispense dopamine, a neurotransmitter that brings on sensations of joy and delight. However, when nothing too exciting is going on, the reward centers get a bit stingy with the feel-good chemicals and you're deprived of that giddy rush. At this point, your prefrontal cortex kicks into boredom mode and you are left with the perception that time is passing slowly. One minute can seem like a thousand years.

The treadmill, in a sense, is the ultimate boredom machine. With no new or interesting stimulus, those neural reward centers have no excuse to release a shot of joy juice. Your feet are plenty busy, but your brain is left desperately scanning its surroundings for something, anything, to latch on to.

While all this is true, it's equally true that most runners view the treadmill as at least an occasional, necessary evil. Sometimes, stepping outside and striding forth simply isn't convenient or safe. Inclement weather, late nights, injury, or any number of other factors may force you indoors and onto the dreadmill. We have certainly covered a lot of this ground in previous chapters.

BOREDOM BUSTERS

I'm here to offer you some strategies for keeping boredom at bay. Try them all. If one doesn't work, another one might. Or you may find one strategy works for a while before soon enough you're transported back to

Dullsville. My advice is to mix, match, and rotate different boredom-busting strategies as often as it takes to keep the brain sparking. You might even find one or two that help you actually enjoy your trip to nowhere.

What if they all fail? I hope that doesn't happen to you. But if it does, perhaps you will find some comfort in knowing that, like doing the laundry or getting your teeth cleaned, a treadmill workout is one of those things you will never learn to love, but, once it's over, you will be glad you did it.

Distraction Strategy #1: **Entertainment**

Almost any gym you walk into has a TV planted squarely in front of each treadmill (and all the other cardio machines, too). If they didn't, they probably wouldn't have any members nowadays. When you use the treadmill you need something to occupy your mind. (Although you should be careful not to occupy it so much you become the subject of one of those epic gym-fail videos that seem to go viral every week or two.)

As you may recall from Chapter 6, diverting your attention with some form of entertainment is an external dissociation strategy. This means that TV, as well as movies and music, help you pass the time by engaging your brain in, say, the news or your playlist, so you can try to forget that you're running in place.

Runners tell me treadmill entertainment works well enough as a diversion—but only up to a point. A lot of them say that TV or music alone isn't enough to escape the mind-numbing boredom of treading, so they sometimes play games based on what they're watching or listening to. For example, marathon runner Ted Curry likes action flicks. He told me he does sprints every time a chase scene comes on the screen, for an average of about six 30-second sprints an hour. Natalie Bell, a casual runner I know, says she tends to trot along while viewing a sitcom but steps up the pace during commercials.

I think these games are a smart idea. Entertainment may not hold your interest long enough to finish a workout. Even if it does, it has been my observation that people tend to go at a much slower pace than normal when they're glued to the tube. So creating an interval workout based on what you're watching or listening to adds an extra layer of engagement on top of the entertainment—and ensures you get a better workout.

Distraction Strategy #2: **Happiness in Numbers**

I'm big on joining support groups to get you through the tough times. Running on the treadmill is no exception.

What do I mean by a treadmill support group? You take what is essentially a solitary activity and turn it into a communal effort. I go into a lot more detail about all the reasons training partners and running groups can help you up your game in Chapter 10, but in this particular instance I'm narrowing the discussion to how finding a partner or a group can make the treadmill workout a more tolerable and possibly even productive experience.

You can invite a friend to work out on the treadmill next to you, which is really no different than asking her to join you for a long Saturday morning run. Or, in one of the more creative uses of treadmills I've seen, you can join a treadmill class, which is sort of like a spin class but on a treadmill. The classes I've seen usually have anywhere from 4 to 15 people set up on consecutive treadmills with an instructor giving orders through a wireless headset.

There's a lot to like about treadmill classes: A coach takes you through a structured workout meant to focus on improving some aspect of your running such as form, speed, or pacing. You're so busy paying attention to the instruction and when to press the right buttons on the console, you forget how dull treadmill running is supposed to be. You also get to work out alongside runners of all shapes, sizes, and abilities, something I personally find very inspiring.

If your gym doesn't have treadmill classes—quite likely, since they haven't caught on yet in the same way spinning or even group rowing classes have—ask management to consider starting one. Or gather a group of likeminded treadmill users together, find a time when there's a good chance you can get a block of treadmills together, and start your own.

Distraction Strategy #3: **Run with Purpose**

Any activity that seems meaningless isn't likely to be stimulating. So if you hop on the treadmill and clip-clop along at the same pace for an hour, you're essentially running right into the arms of boredom. Meaningful minutes are faster than pointless minutes. That's a fact. I recommend setting a goal for each treadmill session and then creating a workout to meet that goal.

For example, if you're looking to up your calorie burn, do an interval workout where you alternate periods of high intensity with periods of low-intensity "active rest." If you want to get stronger, do a hill workout. If you want to work on pacing, do a tempo run.

My theory is that button-pushing activity necessary to change speed and incline is in and of itself an anti-boredom tactic. Button presses are essentially a series of discrete, novel events. You get so wrapped up in the anticipation of the next button push, your mind stays hooked on that. Instead of watching the clock and wondering when 45 minutes has passed, you only have to get through the interval between each button press. And then you reboot your brain and do it again. Even if my theory is just a theory, an overwhelming number of runners I have spoken with about treadmill running rely on this strategy to get through a treadmill workout.

"Running on a treadmill, for me, is pure agony," admits Courtney Vanderfield, a 35-mile-a-week runner who uses the treadmill when it's too hot to run outside. "I can only tolerate it by doing interval training so that it keeps me engaged and breaks it up into lots of small chunks, instead of one long, awful, mind-numbing event."

Jennifer Hastings, a masters sub-elite runner, says she has learned to love the treadmill—yes, love it—by using her time to become a better runner. She says that sometimes she actually prefers the treadmill to outdoors.

"I find that it is easier to work hard on the treadmill. I love to run outside too, but at times the visual distraction and environmental factors like wind and elevation gain interfere with the purity of a simple, hard workout. As an injury-prone runner, I am also grateful for the opportunity to run at times when I am reluctant to get stuck outside somewhere. When I am on the treadmill going fast, I feel like I am flying."

Go figure.

Distraction Strategy #4: **Mix It Up**

Sometimes you just can't do it. That's okay, we've all been there. However, even when the thought of treadmilling is too much to bear, you might not want to give up on it completely.

Try rotating between the treadmill and a few other cardio machines every 5 minutes. You can start out on the treadmill, hop to the bike, then dash to the elliptical and finish up on the rower. There—you've just done 20 minutes of cardio. Now repeat as needed until you complete your allotted workout time.

This helps keep your brain busy much the same way as button pushing does—a discrete, novel event; repeat, repeat, repeat. Before you have a chance to get bored, you're on to the next thing.

Another mix-and-match workout runners tell me they love is treadmill circuit training. That's where you do a minute on the treadmill, jump off, do a set of weights, jump back on, do a set of weights . . . and so on until you finish a complete workout. It's a nice way to multitask cardio and strength training, but just be sure you're comfortable hopping on and off the mill while the belt is moving—and make sure everyone around you knows the machine is still on when you step away to do your weight intervals.

Distraction Strategy #5: **Think Up Something Great**

Yes, jogging on the treadmill is the cardiovascular equivalent to watching paint dry. So what? Perhaps it's time to stop thinking about boredom as a bad thing and reframe it as a valuable mental opportunity. Between smartphones, TV, the Internet, and countless other forms of distraction, our 21st-century brains are so used to being switched on and engaged it's easy to feel panicked anytime there is nothing entertaining going on. Emerging science is beginning to show that even though the brain may give the impression that it powers down when it's bored, it is far from quiet during those dull times.

Studies done at Washington University in St. Louis, Missouri, were the first to identify a neural circuit called the default network, which switches on when the brain is not preoccupied with a stimulus in the external environment. In particular, there seems to be an elaborate electrical conversation taking place between the front and rear parts of the brain, as the medial prefrontal cortex fires in sync with areas like the posterior cingulate and precuneus.[2]

Scientists speculate that this chatter between seemingly disparate parts of the brain could be its way of connecting unrelated ideas. Some interesting brain scan studies suggest daydreaming and creativity are generated by this default network. When the brain is feeling bored, it appears to leave the real world behind and travel to an imaginary place where it can spin stories and rehearse alternate realities.

This is an ideal state for the brain to be in when it comes to problem solving and flashes of genius. So you could try using your time on the treadmill to think up some really good stuff. Try to get through the first few minutes. Admittedly, these can seem lethal. But if you can make it through to the other side, you may find that truly special and unique place where the mind meanders happily along and stumbles across its best ideas.

WRAP-UP

The treadmill may have started out as a form of punishment, but you can learn to love it. In fact, the treadmill may be your only alternative on days you can't, for whatever reason, get out on the road. Learn to embrace your race to nowhere. As I've shown you in this chapter, there are plenty of ways to occupy your mind and pass the time.

PART 5

RESOURCES FOR RUNNERS

IN THE FOLLOWING chapters you will find some terrific resources to help you take all the information you have learned and put it to good use. Coming up next are several worksheets to help you quite literally get your thoughts together, as well as the 7-Step Fit Brain Training Plan that will help you in both training and racing. I finish up with a few lessons from some of the greatest runners of our time. The research shows that elite runners think differently than the rest of us. They will share in their own words what they believe is essential for every runner to know.

CHAPTER 18

Worksheets for a Healthy Mind and Body

"ACT AS IF" VISUALIZATION WORKSHEET

Often, picturing a detailed image in your head can help you understand a specific circumstance or goal more clearly. In the space provided, pick something you are working toward and imagine all the most important details about it. You can then use each of these details separately or together in your visualization practice.

Detail:

____

____

____

Image:

____

____

____

How this image will help me succeed:

____

____

____

Detail:

____

____

____

Image:

____

____

____

How this image will help me succeed:

____

____

____

GOAL-SETTING WORKSHEET

State your BEST goal in detail:

G: Describe three emotions the goal makes you feel.

1.

2.

3.

O: List three measurable aspects of the goal.

1.

2.

3.

A: List three ways the goal will challenge you.

1.

2.

3.

L: List three ways the goal will help you learn and improve.

1.

2.

3.

State your GREAT goal:

____

____

____

____

State your GOOD goal:

____

____

____

____

NEWS FLASH VISUALIZATION EXERCISE

Imagine there is a news reporter from your hometown at your next workout or race. (This can be either an event you have actually run or one you are planning to run in the future.) You have run exceptionally well and before you hit the showers, the reporter hurries up to meet you to ask you about your effort.

Create the news headline that will accompany the "article" about your run. Then, write the article describing what happened, using all the details you can think of—what happened, the sights, the sounds, the feelings, even quotes you gave to the reporter.

Pull out this "clipping" whenever you need a boost or to assist with your mental planning for a future event.

The Headline:

____

____

____

The Article:

____

____

____

____

____

____

____

____

____

OVERCOMING OBSTACLES

As a runner, you will encounter obstacles. Your success depends upon how you tackle them.

Use this worksheet to help you identify and strategically overcome obstacles you encounter. It's a pretty straightforward process. First, describe your obstacles in great detail. Next, list as many alternatives and solutions as you can think of. Finally, create an action plan to put your solutions into play.

Obstacles:

____

____

____

____

____

____

Alternatives:

____

____

____

____

____

____

Action Plan:

____

____

____

____

____

____

PRE-RACE "PEACE OF MIND" CHECKLIST

Use this checklist to ensure you've got everything you need for before, during, and after a race. If you are prone to the pre-race jitters, you will find this process comforting.

BEFORE/DURING THE RACE

- Watch or GPS
- Heart rate monitor
- Shoes
- Running top
- Alternate running top/jacket (in case of weather)
- Sports bra
- Shorts/tights
- Extra leg covering (in case of weather)
- Old sweatshirt or T-shirt (to toss after warm-up)
- Race socks
- Gels/sports drinks
- Hat/gloves/neck covering
- Sunglasses
- Sunscreen
- Water/water bottle
- Band-Aids/NipGuards
- Extra cash
- Music player/phone
- Fanny pack/carryall
- Emergency numbers/personal info
- Race chip/tag/entry info

POST-RACE

- Extra socks
- Extra shirt/shorts/pants
- Waterproof jacket
- Extra shoes
- Comb/brush
- Energy bar/snacks
- Towel
- Aspirin or other pain reliever
- Plastic bag (for ice, dirty clothes, etc.)

RUNNER'S THINKING FLIP CHART

This flip chart is a great reference and training guide for changing negative thoughts to positive thoughts. As I frequently say, "Negative thinking makes your shoes heavy." So on this chart, you can identify a common negative thought that you have and "flip" it to the positive side, using that postive thought rather than the negative thought that is unhelpful to you. Once you get the hang of it, you should be able to flip other negative thoughts that aren't even listed here.

NEGATIVE SELF-TALK	POSITIVE THINKING
I've never done this before.	I have the opportunity to try something new and learn from it.
It's too hard.	I'll tackle it one step at a time.
I don't have the ability.	Training will help me improve.
I'll never find the time.	I'll make it a priority and find time in my schedule.
There's no way I can do this.	I'm going to give it my best.
It's too big a goal.	It's important to me so I'll make it work.
I'm not going to get any better or faster.	I'll keep trying and giving my best effort.
I've failed before.	This is a fresh start.
I'm really tired.	I can gather my energy, relax, and get to the next milestone.
It's too far.	Let's take it one mile at a time.

POSITIVE THINKING WORKSHEET

Your brain's RAS knows: You are what you believe. Help rid your mind of negative thoughts, which often are automatic, ingrained, and subconscious. First, banish negative thinking by noting when these thoughts are the strongest. Write them down. Next, fill in the rest of the columns of this worksheet below

SITUATION	AUTOMATIC NEGATIVE THOUGHT	EFFECT ON PERFORMANCE	

to "replace" them with more positive thoughts. Now every time you have a negative thought, refer to your new affirmative statements to consciously replace the old ones. You can write your affirmations on slips of paper, in your workout log, or even on your arm to keep them handy for when you need them.

	CHANGE YOU WANT TO SEE	REPLACEMENT POSITIVE THOUGHT

CHAPTER 19

7-Step Fit Brain Training Plan

Let's put it all together

THROUGHOUT THIS BOOK I have given you a lot to think about, a lot to work on, and a lot to change. I have supplied concrete to-dos that you can try on the fly. In this chapter, I put together a plan that condenses a lot of what you have learned into one comprehensive plan. While it doesn't contain every tip and trick in the book, it gives you a structured step-by-step program to develop that all-important Runner's Brain.

Adopt this plan for a minimum of 3 weeks. It incorporates many of the theories we have talked about; most notably, the concept of neuroplasticity, which is the ability your brain has to reshape and change. For your brain to reshape and morph into the valuable asset you need as a runner, you've got to engage it. You will find exercises to do while you're running and during nontraining time as well. Both are calls to action to your brain.

THE BRAIN TRAINING PLAN

I ask you to try this plan for at least 3 weeks because that's how long it takes to make some lasting changes to your thinking and the physical structure of your brain. If you adopt this plan for a longer period of time, you will engrave these ideas into your brain in an even deeper way. You can also be creative and develop other activities based on anything else you have learned in the book. This 7-Step Fit Brain Training Plan gives you a head start on cultivating, maintaining, and enjoying your newly minted brain habits across the miles. Everywhere you run, your brain is with you, so take care of this precious, special gear like you never have before.

Day 1:
Identity

GOAL: Strengthening your runner's identity means reshaping the reticular activating system, or RAS, the part of the brain responsible for holding on to your belief system. The more you work on this, the more strongly you identify as a runner and the more confidence you build for training and racing.

TYPE OF TASK: Nonrunning.

MISSION: Whether you are a newbie or grew wings on your feet long ago, filling your world with reminders of running and yourself as a runner is important for building self-assurance and sense of self. It may cost you a few bucks here and there to get this one done, but it is an investment worth making because you'll get a nice return on that investment. Remember that your brain's RAS picks up cues from the environment, regardless of what they are, so you need to fill your surroundings with every indication that you are a runner and part of the running community. Start with the list here, but feel free to get creative.

- Subscribe to *Runner's World* and other running-centric magazines.
- Buy some new gear. Get serious about researching the types of shoes that are available and the technology behind them.
- Go online, perhaps eBay, and purchase some running art, vintage posters, or old race T-shirts to wear. The Salvation Army and Goodwill have

truckloads of T-shirts from charity races, big marathons, and schools. Wash them and wear them even when you're just kicking around running errands or mowing the lawn.

- Read more books about running and autobiographies of runners.
- Subscribe to five running blogs. Consider adding a few thoughts on at least one of them in the comments section. Better yet, start a running blog of your own to share your daily thoughts on running.
- Put a quote from a favorite runner in your e-mail signature. If someone asks you about it, be ready with an answer on why it inspires you.
- Join a running group or start one with some running buddies. Or just invite someone to go running with you.
- Build a running playlist of your favorite pavement-pounding music.
- Without being too obnoxious, mention your daily run in conversation. Don't make it about you, but make it about you and running. If something funny happened, tell that story.
- Watch video clips about running techniques, injuries, or mental strategies. You can start with a video I made about eight must-have strategies for running the Boston Marathon, viewable at my website: DrJeffBrown.com. You will find that the strategies I discuss, for the most part, can be transferred to almost any run you're planning on.
- Gradually change your diet to what a runner would eat. Learn about everything from carb loading to electrolyte replacement, and understand how these things can help or hinder you.
- Just for a goof, adopt a mild runner's superstition: first the right sock, then the left; don't wear the race shirt on race day; eat pasta with no sauce the night before; play specific music for stretching.

Day 2:
Goal Setting

GOAL: Develop objective, measurable, and moderately difficult goals that will in turn shape your thinking and training, and provide an effective feedback loop for self-evaluation.

TYPE OF TASK: Nonrunning.

MISSION: I once gave a talk to runners before the Chicago Marathon, and afterward a man from Southern California came up to me to ask for help. He was running the race with his sister in memory of their father, who had died the previous year from Lou Gehrig's disease. He said he was very concerned that he wouldn't be able to run in the cold temperatures that were predicted, and if he couldn't finish, he felt he would be failing his dad.

He had a clear goal, which is great. But in stating his goal he hadn't accounted for variables outside his control that might prevent him from achieving it. So on the spot, I worked with him to adjust his goal.

First, I asked him: "What would your father say about your sister and you running in the marathon?" He responded with a smile: "He'd say we were crazy for even doing it!" That question literally took the monkey off his back. His dad wouldn't have expected that he would run a marathon, let alone feel disappointed if he didn't finish it. I then told him that his goal of running in honor of his dad was the type of goal that gave him something in common with so many other runners who were there to run the race. Those running in memory of friends lost to cancer or other terminal illnesses; in honor of those who lost their lives on 9/11, whether in New York City, the Pentagon, or the field in Shanksville, Pennsylvania; in memory of the Sandy Hook or Columbine victims; those affected by Hurricane Sandy; or the victims of the Boston Marathon bombing.

Runners run for a multitude of charities, causes, or individuals, I told him. The steps they take benefit not only the runner, but also the countless individuals with unique life circumstances. The honor lies in the time, preparation, expense, and choices it takes to get to the starting line, not the finish line. By putting in that effort for such a selfless reason he had already honored his dad and achieved a pretty big chunk of his goal. I'm pleased to say he understood this and walked away from our conversation with a huge weight lifted off his chest.

This wonderful and loving son at least had a fully formed goal of what he was trying to achieve. So many other runners I have encountered haven't given any consideration whatsoever to this fundamental strategy for success. Sure, you can run from day to day without having stated goals per se, but if you want a sense of purpose and a way to focus, grab a pen or pencil and get cracking on the goal-setting worksheet in Chapter 18. The old saying "Ink it, don't just think it" applies here.

Be sure to share your goals with a few people you trust. That's part of the identity you're building, and people will want to support you in that effort. Also, keep your goals in several prominent places so you have a constant reminder of what you're aiming for.

So now, write down all the goals you can think of, then evaluate each of them using the GOAL system from Chapter 4. If it helps, you can go back to that chapter and review the information in more detail before you start. Here's a quick refresher of the highlights:

- **G:** Can I feel my goal in my **gut**?
- **O:** Is my goal **objective** and measurable?
- **A:** Is my goal challenging yet **achievable**?
- **L:** Will my goal help me **learn** about my running and other abilities?

Day 3:
Positive Self-Talk

GOAL: Recognize the voice you have in your head, and turn up the volume on the positive talk (while turning down the dial on the negative internal chatter).

TYPE OF TASK: Running

MISSION: For your next seven runs—be they short or long workouts—I would like you to be fully aware of the voice in your head. In other words, your self-talk. You can go back and review all the chapters in Part 2 for a ton of ideas on how to do just that, but here I'm going to highlight a couple different methods I haven't suggested before for finding your internal voice. You can pick and choose, do each one on a different day, or do them at different parts of your run.

As you run, say the alphabet to yourself, inside your head, softly and then loudly. Notice how you can adjust the volume. Next, try saying the alphabet backward loudly in your head. Take notice of what you tell yourself when you get frustrated. This little exercise demonstrates how to bring self-talk into the fore of your thinking. Remember, you may need to change the content of your inner voice from negative to positive even when you mess up the alphabet. That's a good lesson on how to adjust your thinking during any type of training or racing setback.

In your head, verbally describe your surroundings in detail. Don't keep it simple—actually, make it as complex and intricate as you can. For example, those aren't just trees you're running past. There are slender birches, 40 feet tall with speckled white bark peeling off the trunks and branches. There are bright green evergreens with thousands of needles spreading out from the center. Notice the leaves and needles on the ground. The smell of the bark and earth. The sound the wind makes blowing through the branches. At the base of one tree is a dog. Again, it's not just a dog; it's a chocolate brown lab with droopy ears, a long, waggy tail, and a red collar; he's pouncing on a slobbery yellow tennis ball that his owner threw from 50 feet away and barking with happiness . . . you get the point.

The rich detail of this thought exercise forces you to use self-talk as you describe things to yourself. You may find that with rich detail you are able to dissociate, too. This is an example of how one type of strategy undergirds another type of strategy. You can get so caught up in the details that you forget that you are running. Another resource you can work into your mental training is the positive self-talk worksheet on page 184. Make plenty of copies and use them liberally until you get the hang of talking to yourself in a positive, internal voice.

Day 4:
Pre-Competition Routine

GOAL: Develop a flexible, prerunning routine to use on training and race days.

TYPE OF TASK: Running.

MISSION: Get some index cards or slips of paper. After your next five runs, write down what you did before, during, and immediately after your run, using a separate index card or paper slip for each run. Take those five index cards and sort through them. You will notice that your routines start to emerge from the information you have collected. Now take a sixth index card and list the things that show up on at least three of the cards. You will notice that you do some activities routinely and others not so much.

Routines are unique to individuals. They may include things like waking up at 6 a.m., eating two scrambled eggs after a run, skipping a shave, answering e-mails, meditating or relaxing for 30 minutes—or some variation on these themes or different themes entirely.

I like how world-class sprinter Michael Johnson used to make sure he got everything done early in the day or the day before so it didn't become a distraction on race day. He liked to relax, do very little, visualize. Thanks to his routine, his entire focus could be on the race.

Once you see a pattern in your prerun behaviors, make a list, laminate it if that helps to memorialize it, and throw it in your bag or backpack so you can reference it when needed. (There's a good one in Chapter 18.) And yes, it's flexible so you can change it when you need to—even lamination isn't forever. But promise me you won't get anxious if you must skip something on your list. You can still perform well and not complete 100 percent of your prerun routine.

Day 5: **Association/Dissociation**

GOAL: Develop the skill of moving your attention and focus either toward or away from your mind and your body. Throughout the course of any run, you will face a variety of circumstances where association or dissociation may come in handy. Association is useful when you need to monitor your body and performance closely, or when you feel like a positive image could be helpful in getting a stronger performance to reveal itself or managing the circumstances you are facing. Dissociation, on the other hand, is helpful when factors in the run aren't so enjoyable: boredom, pain, thirst, or frustration. In a way, you're telling those negative factors to "talk to the hand" when you are dissociating. Remember the slobbery tennis ball in the section about self-talk? You can use that sort of self-talk to help dissociate, or use the many examples provided in Chapter 6. It takes practice, but it's a truly remarkable strategy. It goes without saying—but I'll say it anyway—that if you are experiencing pain or discomfort, your body is sending you an important message you should hear. Never use dissociation in a manner that leads to making things worse for yourself.

TYPE OF TASK: Running.

MISSION: You will like this assignment. Try it several times under different circumstances. First, practice association. Association is when you do a mind meld with your body's movement. You associate your legs with the drivers and cranks of a locomotive that chugs up a hill. In need of relaxation? Associate your forward movement with a sailboat that drifts smoothly in the breeze, or pace your breathing or heart rate with music you sing to yourself, or repeat your favorite mantra, catchphrase, or scripture. All of these are actual associative examples runners have told me about, so they can definitely be helpful.

Now, practice dissociation. Dissociating may come naturally to you, much like driving a car and not remembering the past several miles. Why? Your mind drifted away from what you were doing (for our purposes, running) and focused on something unrelated but important. Try that when you run. Think about your spouse, your kids, your last date, a project at work, or holidays or birthdays that are coming up; design a new clothing ensemble in your head, and buy it later.

There are a host of ways of both associating and dissociating. It can be helpful to read Chapter 6 again to try additional ideas and understand how you have control over this pair of nifty mental strategies.

Day 6:
Visualization and Imagery

GOAL: Use your brain to create images while you run today. You can create images that are relevant to today's run or ones that have nothing to do with it but that, when imagined, give you an advantage in some form. By developing the ability to visualize, you will be able to find multiple uses for it. Michael Johnson used imagery more than 30 times before he raced; many runners visualize themselves performing well, or struggling in a particular way only to visualize how they come back from being knocked down. I use this technique with runners in the medical tent after the running is over. On a cold, rainy day, I may have runners with low core temperatures visualize themselves snuggled up on a sofa with a favorite person or blanket, drinking hot chocolate and watching a blazing fire. Can you hear the fire crackle and pop, too?

TYPE OF TASK: Running.

MISSION: You will recognize, practice, and develop the two key components of visualization: vividness and controllability. Vividness requires that you use all your senses when you visualize, giving your brain as much information as you can that will be useful later on. Controllability lets you create, add, or delete any parts of the visualization.

You can visualize running in the cold, in the rain, up a steep incline, down a hill with shin splints, or kicking it at the finish. It's all up to you. Use this mental strategy, especially on running days. Also, it can be fruitful to use it in the evening or at night after a run, or on a non-running day. Practice is the key. Your brain learns something every time.

As an example, let's take a closer look at process visualization, described in Chapter 5. Remember that with process visualization, you are imagining smaller pieces (the process goals) that will eventually lead to the single big piece, in your case the race (the outcome goal). Before your run today, begin with a chunk of your run—the start. In your car, under a tree, at your desk before you leave work, or somewhere in the grass close to where you will start, visualize as many pieces of the starting point and experience as you can. Let me help you get started by mentioning some things you may notice: the quickness with which you tie your shoes; the sound of the running surface as you grab traction with your first couple steps; the smell of your running clothes if they have or haven't been washed; the tightness of your watch band; the coolness of the air as you breathe in through your nostrils; the strength of the sensation of needing to go to the bathroom; and your positive self-talk about today's run.

There are other factors unique to you that you can also visualize. Remember, this is just one chunk of the run. You can visualize any part of the race, as well as before and after it. You are giving your brain a chance to practice what you would like to accomplish. Seasoned runners can't imagine running without doing some sort of visualization (pun intended).

Day 7:
Relaxation

GOAL: Master the implementation of relaxation before, during, and after running.

TYPE OF TASK: Running.

MISSION: Let's finish up by selecting from the different capacities your brain has to bring about relaxation. My suggestion is to try them all before your run to see which one works best. You can use imagery before a run to relax. Many people imagine relaxing on a beach, with cold saltwater energizing their feet when the small waves come in. Others see themselves in a hammock outside a cabana in a tropical locale. Of course, remember to use all the senses and visualize in real time.

The other quick piece of this mission is four-square breathing, which I described for you in Chapter 7. In today's run, I want you to try four-square breathing well before your run and after it. If you take a break during your run, four-square breathing could be an option to consider if you're feeling tense. I want you to use this fundamental breathing technique because it serves as a perfect example of how strategies for running cross over to strategies for living life. Part of this mission also includes trying four-square breathing at work before a presentation or after a rough call with a client; in bumper-to-bumper traffic; or when you feel overwhelmed with the to-do list at home. Trust me, this little breathing exercise can come in handy in so many places.

WRAP-UP

So there you have it: a head start, literally, on training your brain. Keep in mind that benefits gained in any area of human performance take time. A weight lifter works toward goals, a composer rewrites her scores, and runners must exercise their brains for one of the most mental athletic endeavors one can take on. Give yourself time to learn the basics, then start developing more comprehensive and complex strategies. Also, enjoy the brain benefits of running. Simply knowing that your brain is a healthy space in your body can create enthusiasm and motivation for mastering the Runner's Brain game.

CHAPTER 20

From the Minds of the Greats

Learning how some of the best runners of all time think

IN WRITING THIS BOOK, I was fortunate enough to have meaningful conversations with some of the greatest runners of our time. I picked their brains to find out what—besides their prestigious physical abilities—allowed them to achieve such amazing success in one of the toughest sports there is. Talking to them was fascinating, informative, and highly enlightening. Of course, there is no single personality trait all elite runners have in common. But each has a particular philosophy and lessons to share with everyday runners.

Let me be the first to tell you that you can't copy everything the elites do and expect to become an elite yourself. The research shows that there are some differences in the way world-class runners manage their thinking versus everyone else, even those who are just below the elite level.

The research also shows that sometimes the mental strategies that work for a world-class runner can backfire for the average runner.

All that being true, in speaking with these incredible athletes I found they are extremely proficient at brain strategies that have proven useful for runners at any level. That's what I want to share with you now.

And so, in their own words, I bring you eight of the greats, sharing their thoughts on how having a Runner's Brain has helped them envision success, stay focused, and overcome disappointment. While we can't all be elite athletes, we can certainly learn from their achievements and their mistakes. Here's what they had to say.

JOAN BENOIT SAMUELSON

Joan Benoit Samuelson is running royalty. If you weren't around in 1984—or simply need a shot of inspiration—find the YouTube clip of her emerging from the dark tunnel and onto the track of the Los Angeles Coliseum, her signature white cap catching the sunlight as she strides confidently toward the finish line of the first women's Olympic marathon. That race skyrocketed Samuelson into the limelight. Thirty-plus years later, many consider her the consummate role model of what pragmatic tenacity and hard work can achieve in running.

As a Maine native, Samuelson usually lets the running do the talking for her. Fortunately, she agreed to shed some light on how she sets goals. I think her approach is particularly poignant and should resonate with every runner.

In her own words . . .

"For me, goal setting evolves through storytelling. I try to tell the story especially when it relates to running marathons. My goal setting and storytelling are intertwined.

"In 2008 the Olympic trials were in Boston. I thought, okay, I started my career in Boston with the marathon and now I am 50 years old and Olympic trials are coming to Boston. I will try to qualify for the trials and I'll try to run a sub-2:50. So that was the story. That's what I did.

Blake [Russell], Deena [Kastor], and Magdalena [Lewy Boulet] met me at the finish line when I crossed. They were the three qualifiers for the Olympics that year, and that was a very special moment. I thought I could now walk away from what had been a fulfilling running career.

"Then one day I received a call from Mary Wittenberg, the president of the New York Road Runners Club, asking me if I would come to New York to run the marathon in its 40th year, and the 25th year of my Olympic medal. And I thought, that tells a story—sure, I'll come.

"The same thing happened the following year in the Chicago Marathon. It was the 25th year of the date of my fastest marathon time, and the date was 10/10/10. Well, that was a story I wasn't going to pass up!

"Last year in Boston, before the tragedy unfolded, the goal was to go back to that race 30 years after my fastest Boston marathon time, which was 2:22, and try to run within 30 minutes of that 30 years later. So that was the goal, and that was the story. And then this year we went back to Boston with my son and daughter. It was my son's first marathon. The story I wanted to tell was to try to run within 30 minutes of each other 30 years after the Olympics. I had that goal, but I didn't share that with them. And I did it.

"I haven't come up with the next story yet, so I'm not sure when I'll next run. I think everyone can find the hero in themselves as they start to tell their own stories with whatever it is they are pursuing in running and in life."

JEFF GALLOWAY

If anyone was born to run, it's Jeff Galloway. He started running as a teen and now, nearly five decades later, he hasn't stopped. Along the way he became an All-American collegiate athlete and a member of the 1972 US Olympic Team in the 10,000 meters and an alternate for the Olympic marathon. Even now he is still a pretty competitive masters runner; he says he "only" races one marathon a month, supplemented by a dozen or so races throughout the year.

Galloway tells me he has always thought of himself as an average runner who simply worked hard and thought a lot about what he was doing. In my opinion, Galloway is one of the most cerebral runners I have ever met. He has studied the mind-body connection and successfully applied it to his own efforts as well as those of millions of his fellow athletes through personal coaching, as an editor for *Runner's World,* and with a series of excellent training books.

I asked him how he gets through tough workouts and races. Predictably, he had a well-formed response that tracks pretty closely with what I believe.

In his own words . . .

"I've used three mental training methods for over 40 years with myself and my clients. The first one is mental rehearsal. I visualize the race and the outcome over and over again. It's quite effective for getting you set up for situations that you will probably encounter and then setting up a step-by-step plan.

"But I think mental rehearsal, while valuable, will only get you so far in the race, and then you work with real experience. I use 'magic words' as a form of brainwashing. Magic words distract you from the discomfort while they connect to the spirit hidden inside of you. It's what you put in the forefront of your thoughts that counts. You don't give in to any negative messages you hear in your head when you're under stress. You focus on the positive and maintain control instead.

"Your magic words are based on experiences that occur regularly that shut you down. You find the one or two or three magic words you associate with a successful running experience and as you collect experiences, your conscious brain clicks in and it starts working on finding the same solutions that it did before.

"I'll give you some examples. You can use a magic word like 'relax' and build thoughts around that. You can think to yourself, *There is no pressure on me,* or you could say, 'From the first step, I'm going to relax and enjoy the endorphins' or 'I feel comfortable, supported by all of the energy.' If your magic word is 'power,' you can associate that with phrases like 'I feel good about myself and what I'm doing' or 'I know what I'm doing.' And so on.

"A third mental strategy is one I call 'dirty tricks.' These are quick fixes that you use when you're at the end just to get you from one point to the next. So an example is imagining a giant invisible rubber band—I used to use this one while I was highly competitive. I'd see a person go by and my subconscious would start producing negative thoughts. Then I would get out my giant invisible rubber band, and I would give a little swoosh with my hand and I would throw the rubber band over their head and around their waist and then start cinching them in toward me so that I could pull off their momentum. All of this was fantasy world, but it got me down the road another quarter-mile or half-mile. And then I'd use another dirty trick if needed.

"All three of these strategies keep you in the frontal lobe. They are all designed to shift you into the frontal lobe and the conscious brain. They allow you a complete arsenal of mental tools to get you through almost every running situation."

AMBY BURFOOT

Amby tells me he started running in high school because he was required to do a sport. He chose cross-country because the boys on the team were funny and gossiped about the teachers—and because he could hide out in the woods and slack off.

He quickly found that running wasn't actually so bad. He was such a natural, in fact, that by his junior year he was Connecticut state champ in the 2-mile. That was just the start: In 1968 he won the Boston Marathon. (Fun fact: His college roommates were Jeff Galloway and Bill Rodgers!)

A humble guy, he has always said his biggest accomplishment in running has been to run the annual 5-Mile Road Race on Thanksgiving Day in Manchester, Connecticut, every year since 1964. I think what he has to say about how to think about overcoming setbacks is bang-on.

In his own words . . .

"As a competitive runner I was definitely a goal setter. Most of the time I was setting goals for weekly mileage and pace in my workouts. We followed that stuff really obsessively back then, when everyone was running 100 to 125 miles per week. Of course, once I realized I had some talent and met with some success, I began adjusting my goals and setting ones that people who are successful in the marathon always go for: I wanted to win Boston and I wanted to make it to the Olympic Games. Well, I ended up at 50 percent; I won Boston.

"I live with that. Nobody has a good day every day, every big race. The big mystery in running is, why is that? I knew what the formula for winning was; why didn't I use it again and again? The answer is, I thought I did, but it just didn't reach the same ending. But instead of saying I didn't wind up reaching my goal, I think about the positive aspects. I made it into the Olympic trials and I was the ninth-fastest runner in America.

"So one of the lessons in running is that there is more disappointment than anything else. I think it's one of those things that running teaches you. Once you get to a certain level of high achievement, you are much more likely to lose a race than win it. It is a negative until you turn it around and refuse to let it be. You find enough in it to keep motivated and keep hoping for the return of that great day. I learned from the great running coach Jack Daniels that the great day you have is not a fluke. That's your ability. That's who you are. You don't hit it every day, but that is what you are capable of. I like that. The goal for me was always to figure out the right formula and get back and have that great day again."

DEAN KARNAZES

Dean Karnazes is "The Man Who Can Run Forever." As someone who can push the pace for 3 days straight on very little rest, his lactate threshold seems limitless. While his ability to run hundreds of miles without tiring is a genetic gift, he attributes a large part of his abilities to psychology. You may not want to adopt his approach to dealing with pain, but the advice Karnazes offers about mindfulness and staying in the moment is something every runner can learn from.

In his own words . . .

"There's really not much I don't like when I'm running. I've run through mud, in water, sand, and snow, and over rock and none of it really bothers me. When I engage in any physical conquest, such as running a marathon or an ultramarathon, I go into it with the simple commitment to myself that I will try my hardest and give it my all. You can't control the weather, you can't control the other competitors, and you can't foresee the unforeseeable. No matter what, the commitment to be the best me that I can be is unchanged.

"The other thing I have done is shifted my paradigm in respect to pain and struggle. Instead of trying to avoid it, I welcome the hurt and celebrate it. Bring it, baby! *You just try to break me, Mr. Pain,* I say in my mind—*you're no match for my resolve!*

"In times of great doubt and uncertainty as to whether I can continue, instead of chanting a mantra or trying to occupy my mind with superfluous thoughts and banter, I focus instead on being present in the here and now. Rarely do we live in the now. Our minds are constantly thinking about the future or reflecting upon the past. We occupy our thoughts with an unending stream of internal dialog. In times of great duress, I do none of this. I focus only on the present moment of time and the process of putting one foot in front of the other to the best of my ability. There is nothing else going on in my mind other than focusing on taking my next footstep as best as I possibly can. Using this technique has gotten me through unimaginable low points."

KATHRINE SWITZER

I am so pleased to include Kathrine Switzer's words in this book. Not only is she the first woman to officially run the Boston Marathon and a subsequent winner of the New York City Marathon, she is a feminist icon who broke down walls and fought for every woman's opportunity in sport. It is one thing to win big races, and another to blaze a trail.

Switzer is perpetually upbeat. Here she shares a few words about how to stay sane and positive while working through an injury.

In her own words . . .

"When I can't run, like when I'm injured, I do something else that's fulfilling, like writing a book, launching a project, moving ahead in my career, re-mortgaging the house, cleaning the attic. Then I use the training time to do other things that help keep me fit, too—swimming, stretching, yoga, weight work—all things I didn't have time to do while running competitively.

"My advice to other runners: When it hurts, don't run on it. Give it lots of time to heal and you'll come back faster and stronger, refreshed and eager. The idea that you will lose all that fitness entirely is wrong. You retain a lot. Relax. Don't panic. Spend some of that time paying attention to people you love and you often don't have as much time for when you're training."

FRANCIE LARRIEU SMITH

Francie Larrieu Smith, a five-time Olympian, is known as one of the toughest and most resilient track and field competitors of all time. Along with women like Samuelson and Switzer, she has inspired a generation of female runners. She now coaches runners at the university level.

Smith is humble, hardworking, and honest. When she speaks about running, she is incredibly pragmatic. Yet I find her thoughts on visualization vivid and imaginative. In describing her use of this mental strategy, I think she perfectly captures what I have told you about the importance of visualization in Chapter 6.

In her own words . . .

"I would use visualization before races all the time. The best example that I can give you was when I raced indoors. Indoor track is a very crazy environment and very small, and I loved it so much. But when I would go out to the infield during the last minutes of prep, I would be in the middle of the track, and you've got races going on around you and the crowd is up on their feet because the race is going on. So you really need to kind of turn everything inward and focus on the race. And so how I did that in that crazy environment was, I would look up into the rafters of the building, and you see those lights, and I would just focus on that light and think about the race. I would see myself running the race and winning the race.

"To me it's just turning all of your energy inward and focusing on the event. Now that I work with athletes as a coach, I know that applies to runners of all levels. I encounter people who are terrified of the starting line because they have a case of the nerves. I try to explain to them that that's normal and you want that. It's just learning how to channel the energy, to spend some time on picturing how you want the race to go.

"Everybody gets nervous going to the starting line. But I think when you are in that state, it's important to try to think inward. Visualize what it is you're going to do. Maybe close your eyes and just think about the race. Then get up, do a couple hard strides, and then pretty soon you're on the line."

MEB KEFLEZIGHI

As the greatest marathoner in America, Keflezighi has done it all. He's won Boston, New York, and a silver medal in the Olympics.

These are pretty big accomplishments for a skinny kid who came to America as an Eritrean refuge at the age of 12. You would think he has only had success. But the perpetually optimistic Keflezighi has had injuries, setbacks, and, frankly, some pretty crummy races. It never stops him. Here he recounts what happened when he hit the wall in the 2013 New York City Marathon. I think you will find his ability to think about his so-called failure as a kind of victory—and unbelievably motivating. Perhaps it will help you reframe your thinking about what it means to be successful as a runner.

In his own words . . .

"At the 2013 New York City Marathon, I lacked training because I was coming back from an injury and my body was conditioned to go 18 or 19 miles. And it was not ideal preparation. But I pushed my body to the maximum and I was good for half. And then at about 15, the other leaders picked it up. I couldn't go with them. I pushed it for four miles and then said, *Uh-uh, you're done.*

"Before about 16 miles I was under a 5-minute [pace], and then—that ache! I ran from 19 miles to 19.3 miles and then I stopped. I walked, jogged, walked, jogged the rest of the 0.7 miles, and that split was 9:58—it was twice as long.

"And now what do you do? Do you stop and get in the van? But I decided, you know what? Just because of what happened with the bombing in Boston and because of what happened with the cancellation of the New York City Marathon the year before, maybe I'm not going to win the race today but I'm going to find a way to get to that finish line. And that's what I did.

"Many people encouraged me to keep going with them, and I just couldn't. And then when [amateur runner] Mike Cassidy caught up to me with 3 miles to go, I said, well, I'll try. And then we just basically helped each other out. And I said, you know what? We're going to hold hands and finish together. So it isn't always about winning, but getting the best out of yourself."

MICHAEL JOHNSON

"Life is often compared to a marathon, but I think it is more like being a sprinter: long stretches of hard work punctuated by brief moments in which we are given the opportunity to perform at our best," says Michael Johnson, one of the greatest sprinters of all time. His four Olympic gold medals, eight World Championships and two world records are just the highlights of his career. We would be here all day if I ran down every single one of his accomplishments.

With all of Johnson's street cred, this chapter wouldn't be complete without his advice on visualizing success. Predictably, it's short, fast, and right on track.

In his own words . . .

"I'm a very goal-oriented person. It took me some time to realize how important my mental training actually was. Once I developed the skill, I thrived on the consistency of brain training being in my physical training routine.

"There doesn't have to be something wrong with you to work on strengthening your mental skills or learning new ones. I developed skills to help me stay focused and to deal with heavy pressure. I did that by eliminating distractions on race day so I can control my environment and thoughts better."

NOTES

CHAPTER 2

1 Alison Chan, Jennifer C. Davis, Lindsay S. Nagamatsu, et al., "Physical Activity Improves Verbal and Spatial Memory in Older Adults with Probable Mild Cognitive Impairment: A 6-Month Randomized Controlled Trial," *Journal of Aging Research* 2013 (2013), doi:10.1155/2013/861893.

2 F.H. Gage, T. Shubert, H. van Praag, and C. Zhao, "Exercise Enhances Learning and Hippocampal Neurogenesis in Aged Mice," *The Journal of Neuroscience* 25, no. 38 (2005), 8680–8685.

3 M.A. Babyak, J.A. Blumenthal, K.A. Moore et al., "Effects of Exercise Training on Older Patients with Major Depression," *JAMA Internal Medicine* 159, no. 19 (1999), 2349–2356.

CHAPTER 4

1 Gary P. Latham, Edwin A. Locke, Lise M. Saari, and Karyll N. Shaw, "Goal Setting and Task Performance: 1969–1980," *Psychological Bulletin* 90, no. 1 (July 1981), 125–152.

CHAPTER 5

1 Jing Z. Liu, Vinoth K. Ranganathan, Vinod Sahgal, Vlodek Siemionow, and Guang H. Yue, "From Mental Power to Muscle Power—Gaining Strength by Using the Mind," *Neuropsychologia* 42, no. 7 (2004), 944–956.

2 Alfredo Campos and Maria Jose Perez, "Vividness of Movement Imagery Questionnaire: Relations with Other Measures of Mental Imagery," *Perceptual and Motor Skills* 67, no. 2 (1988), 607–610.

3 Charles A. Garfield, *Peak Performance*, Warner Books (March 1987).

CHAPTER 6

1 W.P. Morgan and M.L. Pollock, "Psychologic Characterization of the Elite Distance Runner," *Annals of the New York Academy of Sciences* 301, October 1977, 382–403.

CHAPTER 7

1 A. Berthele, H. Boecker, G. Henriksen, et al., "The Runner's High: Opioidergic Mechanisms in the Human Brain," *Cerebral Cortex* 18, no. 11 (2008), 2523–2531.

2 A. Dietrich and W. McDaniel, "Endocannabinoids and Exercise," *British Journal of Sports Medicine* 38, no. 5 (2004), 536–541.

3 Mihaly Csikszentmihalyi, *The Psychology of Optimal Experience*, Harper Perennial Modern Classics (1990).

4 E.E.A. Cohen, R. Ejsmond-Frey, N. Knight, and R.I.M. Dunbar, "Rowers' high: Behavioural synchrony is correlated with elevated pain thresholds," *Biology Letters* 2010; 6(1):106–108.

CHAPTER 8

1 B.F. Skinner, "'Superstition' in the Pigeon," *Journal of Experimental Psychology* 38, no. 2 (1948), 168–172.
2 P. Brugger and C. Mohr, "The Paranormal Mind: How the Study of Anomalous Experiences and Beliefs May Inform Cognitive Neuroscience," *Cortex* 44, no. 10 (Nov/Dec 2008), 1291–1298.
3 D. Castro, J. Fossella, T. Hines, and A. Raz, "Paranormal Experience and the COMT Dopaminergic Gene: A Preliminary Attempt to Associate Phenotype with Genotype Using an Underlying Brain Theory," *Cortex* 44, no. 10 (Nov/Dec 2008), 1336–1341.
4 Christine Hosey, Jane L. Risen, and Yan Zhang, "Reversing One's Fortune by Pushing Away Bad Luck," *Journal of Experimental Psychology: General* 143, no. 3 (June 2014), 1171–1184.
5 L. Damisch, T. Mussweiler, and B. Stoberock, "Keep Your Fingers Crossed!: How Superstition Improves Performance," *Psychological Science* 21, no. 7 (July 2010), 1014–1020.

CHAPTER 9

1 Adam D. Galinsky and Adam Hajo, "Enclothed Cognition," *Journal of Experimental Social Psychology* 48, no. 4 (July 2012), 918–925.
2 Yoon-Hee Kwon, "The Influence of the Perception of Mood and Self-Consciousness on the Selection of Clothing," *Clothing & Textiles Research Journal* 9, no. 4 (1991), 41–46.
3 B. Fletcher, N. Howlett, I. Orakcioglu, and K. Pine, "The Influence of Clothing on First Impressions: Rapid and Positive Responses to Minor Changes in Male Attire," *Journal of Fashion Marketing and Management* 17, no. 1 (2013), 38–48.

CHAPTER 10

1 Norman Triplett, "The Dynamogenic Factors in Pacemaking and Competition," *American Journal of Psychology* 9, no. 4 (1898), 507–533.
2 J.C. Eisenmann, D.L Feltz, B.C. Irwin, N.L. Kerr, and J. Scorniaenchi, "Aerobic Exercise Is Promoted When Individual Performance Affects the Group: A Test of the Kohler Motivation Gain Effect," *Annals of Behavioral Medicine* 44, no. 2 (Oct 2012), 151–159.
3 D.L. Feltz, B.C. Irwin, K.A. Osborn, and N.J. Skogsberg, "The Köhler Effect: Motivation Gains and Losses in Real Sports Groups," *Sport, Exercise, and Performance Psychology* 1, no. 4 (Nov 2012), 242–253.

CHAPTER 14

1 J.A. Dempsey, H.C. Haverkamp, J.D. Miller, D.F. Pegelow, and L.M. Romer, "Inspiratory Muscles Do Not Limit Maximal Incremental Exercise Performance in Healthy Subjects," *Respiratory Physiology & Neurobiology* 156, no. 3 (June 2007), 353–361.
2 Victoria Manning, Samuele M. Marcora, and Walter Staiano, "Mental Fatigue Impairs Physical Performance in Humans," *Journal of Applied Physiology* 106, no. 3 (March 2009), 857–864.

3 E.-J.M. Fares and B. Kayser, "Carbohydrate Mouth Rinse Effects on Exercise Capacity in Pre- and Postprandial States," *Journal of Nutrition and Metabolism* 2011, doi:10.1155/2011/385962.

4 S.J. Biddle and C.D. Stevinson, "Cognitive Orientations in Marathon Running and 'Hitting the Wall,'" *British Journal of Sports Medicine* 32, no. 3 (1998), 229–235.

CHAPTER 17

1 Anthony Vaver, "Prisons and Punishments: The Failure of the Treadmill in America," Early American Crime, www.earlyamericancrime.com/prisons-and-punishments/failure-of-the-treadmill.

2 Wei Gao, John H. Gilmore, Kelly S. Giovanello, Weili Lin, Dinggang Shen, J. Keith Smith, and Hongtu Zhu, "Evidence on the Emergence of the Brain's Default Network from 2-Week-Old to 2-Year-Old Healthy Pediatric Subjects," *Proceedings of the National Academy of Sciences* 106, no. 16 (2008), 6790–6795.

ACKNOWLEDGMENTS

ALTHOUGH THE NOTES reference much of the scientific research used to support the ideas offered in *The Runner's Brain,* they do not begin to express the gratitude I feel for all of the researchers and their subjects who contribute to what we know about the brain and running. Without their scientific exploration much of what we know to be true would remain pure speculation or hidden from view.

First, no team member I've ever worked with is stronger or more resilient than Liz Neporent. She comes from behind, lifts the heavy weights, hurdles obstacles with grace, and still has time for brainstorming, laughing, and taking life in stride. A special thanks to Jay and Skylar Shafran, Liz's family of athletes, for graciously providing Liz with both writing and running time without complaint. The two of you are the best cheerleaders Liz could ask for!

Many thanks to our tough-as-nails agent Linda Konner for guiding this book through from concept to completion. And just as many thanks to our patient and skilled editor at Rodale, Mark Weinstein, who has made *The Runner's Brain* even more useful to readers.

Kudos, gratitude, or any other nifty word that describes indebtedness goes to *Runner's World* editor, Katie Neitz, who values runners' brains and has been supportive of *The Runner's Brain* from its inception. She and the entire *Runner's World* magazine, website, and social media crew were available on demand all through the writing process.

Katie Moisse, Leigh Devine, and Janet Ungless were also kind enough to act as sounding boards. Additionally, my go-to posse of Gretchen Brown, Nancy Scott, Rusty Shelton, Chris Troyanos, Mike and Meg Greto, Michael and Victoria Landers, Stephen and Jill Shoemaker, Chris and Jodie Huff, Nate and Nicole Jui, Beth Meister, Arthur Siegel, Nathan and Brea Ashcraft, Konrad Osa, Lisa Miller, Stephen and Kathryn Hubbell, and all of my Boston Marathon and Boston Athletic Association colleagues and friends are exceptional and priceless.

Finally, we wish to express gratitude to the 65 million runners in this country who take to the streets, tracks, and treadmills on a regular basis. You inspire us every day with your dedication and hard work. Whether your focus is a local fun run or Olympic trials, keep on logging those miles!

ABOUT THE AUTHORS

Dr. Jeff Brown is the lead psychologist for the Boston Marathon medical team and is an assistant clinical professor in the department of psychiatry at Harvard Medical School and McLean Hospital. A bestselling author, Dr. Brown is a pioneer in taking psychology and cognitive behavioral psychology to the masses. He serves on *Runner's World* magazine's scientific advisory board and is frequently cited in major media outlets. He knows astutely how runners think and behave, whether it's on a run through the neighborhood or stepping up to the starting line in honor of someone special or returning to run a race interrupted by the 2013 Boston Bombings. Dr. Brown is passionate about educating and encouraging athletes who have an insatiable desire to grasp their goals by maximizing individual brain potential. Among his many quotes, he is known for saying, "Negative thinking makes your running shoes heavy." Learn more about Dr. Brown at drjeffbrown.com.

Liz Neporent is a health and fitness expert, writer, and social media consultant. She is former health reporter for ABC News National and author and coauthor of 15 health books including the bestseller *Fitness for Dummies,* now in its 4th edition. She also serves on the emeritus board and is a national spokesperson for the American Council on Exercise, a leading national authority on fitness and weight loss. Liz is a lifelong "pathological" runner, having run 25 marathons, 6 ultra marathons, and countless road races at a variety of distances. In fact, when she's not with her husband, Jay, and daughter, Skylar, she can usually be found in the gym or on the road. Connect with Liz on Twitter @Lizzyfit.

INDEX

<u>Underscored</u> page references indicate boxed text.

I

J

K

L

M

N

Q

R

S